STEEL
BOAT
BUILDING

STEEL
BOAT
BUILDING
From bare hull to launching

$\triangledown$

Volume 2

Thomas E. Colvin

INTERNATIONAL MARINE
PUBLISHING COMPANY
Camden, Maine

Typeset by The Key Word, Inc., Belchertown, Massachusetts
Printed and bound by BookCrafters, Chelsea, Michigan

Published by International Marine Publishing Company
21 Elm Street, Camden, Maine 04843
(207) 236-4342

Library of Congress Cataloging in Publication Data

Colvin, Thomas E.
   Steel boatbuilding.
   Includes index.
   1. Steel boats. 2. Boatbuilding. I. Title
VM321.C68 1985   623.8'207   84-48520
ISBN 0-87742-189-7 (v. 1)
ISBN 0-87742-203-6 (v. 2)

$\triangledown$

CONTENTS

$\triangledown$

PREFACE

When I first began to write a book on steel boatbuilding, I planned to cover the subject in a single volume. After receiving several of the early chapters, however, the publisher asked me to expand the work to encompass steel boatbuilding in its entirety, as far as was practical. Thus, we deemed it advisable to terminate Volume 1 with the chapter on interior welding details, so that that volume covers the basic aspects of constructing a steel hull. Volume 2 continues the building of a vessel to its ultimate end—launched, rigged, and ready for sea. This volume is also apropos for those who already own a steel vessel and may be contemplating maintenance, repairs, or alterations or modifications of its arrangement or rig. There is a thorough chapter on finishing, it being an unarguable fact that the longevity of any vessel, new or old, depends entirely on proper finishing and maintenance.

Inseparable from the building or owning of a vessel is its employment or use. A consideration for the builder is whether to make his avocation a profession, and a consideration for the owner is whether to employ the vessel commercially. These considerations, too, are covered in Volume 2.

Thomas E. Colvin

1

▽

INSULATION, HEATING, AND REFRIGERATION

The habitability of small seagoing vessels is at best not much better than that of a slum tenement; indeed, it is often likened to an overcrowded jail cell. There are laws applying to larger vessels, and among that host of laws are those that dictate the minimum allotment of space for bodies, the sanitary minimums, food quantity and quality, and the hours that a seaman may work. Small vessels, however, can seldom comply with all these laws. Good shipmates help make a small vessel's accommodations acceptable and comfortable, and habitability is improved by ventilation, heating, the circulation of air through interior spaces, and dryness.

Insulation is needed to some extent on all vessels for the comfort of the crew. The amount will depend on the waters in which the vessel sails and the type of heating or cooling equipment that is to be used. Ventilation—and its importance for the vessel as a whole—was mentioned in Chapter 14 of Volume 1, but ventilation is also important in determining the efficiency of a vessel's insulation. In the living accommodations, insulation and ventilation alone are not sufficient; there must also be a circulation of air to all spaces, including lockers, bins, and drawers, with no stagnant compartments or voids.

INSULATION

Insulation aids in the maintenance of temperature only when combined with either heating or cooling devices. In the absence of these devices, the material assumes no value in itself. Selecting the proper insulation material is important, because a good material may also render other functions, including those of flotation, sound absorption, and a fire barrier. The most common insulation materials are cork, charcoal, mica minerals, sawdust, fiberglass, mineral wool, perforated asbestos cement board, polystyrene boards, foamed-in-place polyurethane, polyurethane boards, cellulated glass blocks, and wood. These materials may be used alone or in conjunction with one or more other materials to achieve the required insulation factor, finish, and safety.

The selection of insulation should never be based solely on the cost of material, as is the common practice of house builders. Rather, selection should be predicated on the material's workability, stability, weight, compressive strength, and other properties. Is it combustible? Does it lack any odor? Will it not support vermin, mold, or bacteria? Does it have the lowest possible thermal conductivity? A material may have all of the above yet still not be the primary choice. It might need to be overlaid, causing a great deal of waste; it might occupy too much volume; or there might be extra labor involved in working it.

No insulation should be applied to bare steel with the hope that, once the steel is hidden, no problem will develop. As with everything else that is used aboard a vessel, nothing is perfect. The hull of the vessel should be treated for corrosion just as if it were not to be insulated, with the exception that finish color coats of paint need not be used; however, an additional barrier coat is in order.

In vessels used primarily in the tropics, only the decks and cabins exposed to the weather need insulation. The areas that surround the engine room need a sound barrier more than insulation from heat, since living areas as well as engine rooms will be, for the most part, naturally ventilated, and the engine room is rarely much hotter than the cabins. Except in calms, enough air will circulate for comfort; otherwise, small fans can supplement the natural circulation. Decks and cabins must be painted white to reflect the heat.

As a vessel becomes farther removed from the tropics, the need for hull insulation increases. In temperate zones, the insulation is extended downward to the in-ballast waterline. In subarctic and arctic zones, insulation carried all the way down to the bilge will aid in the retention of heat. If only the living areas are to be insulated, then the insulation must extend below the cabin sole.

Conversely, fish holds need greater amounts of insulation as a vessel nears the tropical waters. The insulation in the tropics needs to be three to five times as efficient as it would have to be in subarctic waters.

Air is one of the most convenient forms of insulation when confined, say, between the shell of the hull and the ceiling. A ceiling is not an integral part of the structure in a steel vessel; therefore, when one is required it can be installed in such a way that it is easily removed, and any future repairs to the hull will require the minimum of rebuilding. The potential disadvantage of using air for insulation is that a tight ceiling will cause a damp, stagnant air space, which in turn will rot the ceiling and accelerate corrosion of the hull. This damp void is also conducive to the harboring of mold, bacteria, and rats and other vermin. Leaving an air course along the ceiling's upper edge, however, permits enough air circulation to virtually eliminate the rot and dampness, thereby reducing the possibility of corrosion.

When space is at a premium, which it usually is in small vessels, there is no room for a ceiling, and other methods must be employed wherein the insulation is in direct contact with the shell of the vessel. The common fault in all these methods is that the insulation must be removed to effect any repairs to the hull, and casual inspection of the shell interior is impossible.

Fiberglass batts are the most common form of insulation in living spaces. The batts are available in 1-, 1½-, 2-, 3-, and 4-inch thicknesses, and in widths of 15, 20, and 30 inches. A thickness of about 1½ inches is used in small vessels due to space limitations. Fiberglass is easily cut to whatever size is required, and is also available in a board form with a fire-resistant binder. The fiberglass boards may also have a fiberglass cloth backing on one side that is either plain or perforated.

Fiberglass can be laid directly against the protected steel; it is hydroscopic, however, and under certain conditions will permit condensation over the interface. The semi-rigid boards with a fiberglass cloth face can be fitted without any fastenings on steel vessels that use T-bars as longitudinals; the boards are flexed crosswise and the edges compressed enough to pop them under the flanges of the T's, which will hold them in place with a snug fit all around. This method of installing semi-rigid boards has been acceptable to the U.S. Public Health Service as a means of ratproofing fiberglass more than 1 inch thick. The flanges of the T's must be covered with a separate material that overlaps the fiberglass boards, because the exposed steel will sweat under the right conditions. The same precaution applies to the transverse frames. Should a builder elect to use the fiberglass batts, he will need to resort to mechanical fasteners that are welded or glued to the hull, plus wire mesh over the fiberglass to hold it in place and make it ratproof.

Fiberglass, with all its virtues, will shake down and out if not fully sheathed, causing skin irritations for some people. The fine particles of glass are a health hazard, particularly if ingested with food.

Mineral wool has a tendency to settle and will retain moisture; therefore, it is not often used on small vessels except when it can be isolated from the hull.

Cork is an excellent insulating material and is available in such a variety of compositions and textures that it can also be used as a decorative finish. The common slabs of granulated and compressed cork are from 1 to 3 inches thick, although any size and thickness may be customized at a very low additional cost. The cork boards may be attached to the shell of the vessel in a variety of ways, the most common being by use of a contact cement or several dabs of quick-setting epoxy adhesive available at local builders' supply houses for the attachment of paneling and wallboard. The cork must fay against the steel in its entirety, and the edges must be sealed airtight, which can be very time-consuming. An alternative is to fit the cork to the shell plating but leave a gap around all structural members; the gaps are then filled with expanded polyurethane foam, which assures a watertight seal. The cork must be at least as thick as the longitudinals. The surplus foam is sanded or rasped off to form a continuous smooth surface. This surface is then overlaid with cork floor tiles that have an adhesive backing, or with plain cork tiles or board using contact cement. The floor tiles are available prefinished, which saves time and labor for the builder.

Expanded polyurethane foam is handled in two ways: it is either poured in place after mixing the two components, or it is shot in place using a special mixing nozzle that combines the two components under pressure at the moment of separation from the nozzle. The latter is most often used to insulate an entire hull or compartment, and has the most uniform as-mixed qualities. Control of the thickness depends on the skill of the operator. Even the most skilled operators cannot obtain an absolutely smooth, even finish on the interior of a vessel. There may be areas where this is unimportant, but in general, while the foaming itself can be accomplished in a minimum of time, the cleanup afterwards to an acceptable uniformity is very time-consuming. In my own vessel of 48 feet 6 inches, about five hours was required for the foaming of the hull (deck edge to keel), the undersides of all decks, and the undersides of the cabintops to a thickness of 2 inches. The cleanup time to an acceptable uniformity was 120 hours of cutting, sanding, and rasping, which removed five drums of surplus material. The cleanup removed the skin that forms during the curing process, leaving a more vulnerable surface. The foam is a fire hazard under certain conditions; therefore, it was overcoated with a fire-retardant paint that expands when heated.

The foam should never be sprayed directly on bare metal because, even though it is advertised as being closed-cell, water can and will enter the surface cells. Also, there is a probability that some cells will not form during the foaming process. The water vapor through foamed urethane is about four times higher in concentration than are atmospheric gases, and where differential temperatures are involved, the water vapor will condense at the foam-metal interface. Freeze-thaw cycles will cause a separation of the bonding, and water vapor tends to carry fluorides and chlorides to the interface, destroying the bond between the steel and the foam. The presence of water at the interface along with oxygen will cause a typical oxygen corrosion cell which, in turn, causes "poultice" corrosion. The corrosion products then expand, causing a separation zone between the steel and the foam. By using the Swift Company's Fome-Bond No. 64, most if not all problems of adhesion will be eliminated.

The poured-in-place foam is often used, as it is a quick way to fill a cavity, especially in the way of iceboxes and refrigerators. The foam is mixed and poured as a liquid. Extreme care is needed when mixing to assure uniformity and the absence of voids created in pouring. Great pressure is exerted during the process of expanding, and this must be restrained with adequate shoring or the bulkhead and box liner will be deformed by bulging. Since considerable heat results from the mixing of the materials, the pours must be limited in volume. This aids in the reduction of pressure except in very small cavities.

Polyurethane planks can be obtained in several densities, and are easier to use in that the surface is smooth and can be overlaid without further preparation. When laying the planks against the shell plating, as with cork, it is best to attach them with contact cement and to gap them all around in way of metal structures (longitudinals and transverse framing). Then, using polyurethane foam, seal all the edges and smooth out the surplus. The surface may then be finished with a wood veneer or cork. (See Figure 56 in Chapter 9 of Volume 1.)

Polyurethane foams are not approved by the United States Coast Guard for use in vessels carrying passengers for hire or in any other vessel subject to Coast Guard certification, unless the foam is fully encased in a fire-retardant structure, because the burning foam gives off cyanide gas.

By covering the foam with any of a variety of materials, it is possible to construct a fire-resistant interior. Even a fireproof interior might be possible, but in small vessels, the result would also be a sterile and depressing arrangement. There is no reason to fear the use of one material or another as long as its use is thought out and understood. Fire is disastrous on any vessel, but the conditions that cause fire can be avoided. A fire is generally caused by dirt and filth, which result from poor shipkeeping, negligence, carelessness, and ignorance.

When total insulation is not a requirement, the proper protective coatings on the steel may be covered by a mixture of either ground cork or small grain mica material, such as muscovite or potash mica. These mixtures can be sprayed on the steel, which roughens its surface, expanding the effective surface area several hundred times beyond that of plain steel. This type of finish will not eliminate condensation, but the condensation will not run down the greatly expanded surface, and so has a better chance of evaporating.

Condensation

Condensation occurs after water vapor is added to the air to the limit of the air's absorption capacity; when the air temperature is lowered below its dew point, the absorption capacity is exceeded. Condensation is further encouraged by the presence of minute salt crystals—primarily sodium, but also magnesium, calcium chloride, and sulfates, all of which are components of seawater. The higher the air temperature, the more water vapor the air will hold, up to a maximum of about three percent. At low temperatures, air will "carry" only about 0.01 percent water vapor. Thus, condensation will always be a greater problem in cold climates than where it is warm.

Thermal Conductivity of Materials

The thermal conductivity of any material is the quantity of heat transmitted per unit of time, per unit of cross section, and per unit of temperature gradient. The best heat conductors are metals; liquids are poor conductors, and gases are almost nonconductors. Therefore, to achieve the same degree of comfort from the various materials used in boat and ship construction, it is necessary to consider their conductivity relative to each other. The common materials used in the building of small vessels can be compared when related to a material selected as the standard. Inasmuch as most of the interior joinerwork in small vessels is wood, and wood is an insulator, one may use it as the standard for comparison. Table 1 summarizes the most common comparisons.

**TABLE 1.   Thermal Conductivities of the Materials
Commonly Used in Small Vessels**

Material	Conductivity cal/cm/sec/°C	Conductivity Relative to Wood
Copper	0.918	5,738.00
Aluminum alloys	0.480	3,000.00
Cast iron	0.161	1,006.00
Lead	0.083	519.00
Ice	0.005	31.25
Glass	0.002	12.50
Water	0.00139	8.69
Asbestos (board)	0.00087	5.44
Wood	0.00016	1.00
Cork	0.001	0.63
Mineral wool	0.00009	0.56
Fiberglass (blanket)	0.00008	0.50
Fiberglass (board)	0.00007	0.44
Polyurethane (foam)	0.00005	0.31
Polyurethane (plank)	0.00005	0.31

HEATING

The major difference between sailing vessels and steam vessels in the habitability of their fo'c's'les was the degree of comfort derived from steam heat, which was in abundant supply on the latter. In some sailing vessels the galley was in the fo'c's'le rather than in a deckhouse, and this provided a greater degree of comfort than when just a space heater was used. In some large sailing vessels, the crew was berthed in a deckhouse, where, under certain conditions, downdrafts made it impossible to keep the stove burning properly, and the fire would have to be extinguished because of the smoke billowing from the firebox. In the summertime, the heat from a cookstove coupled with poor ventilation made a below-deck fo'c's'le a hothouse, so sleeping on deck under a tarp was common. Windsails (scoops) were rigged as often as possible in order to keep a reasonable amount of air circulating; combined with the dry heat of the stove, this kept the fo'c's'le from becoming a total mushroom factory.

In vessels with cabins that are separated by a cargo hold, it is necessary either to duplicate the heating system or to insulate the heating delivery vehicle through the cargo hold. Even in yachts, there will be some difficulties in heating the complete interior with just a stove or space heater. Cabins are heated with hot water baseboards or radiators, forced draft (warm air) heat, convection (radiant) heat, or a combination of these.

The prerequisite of any heat source is that it must be vented to the exterior of the vessel, and the space in which it is used must also be ventilated. Regardless of the type of fuel used, there is always a loss of oxygen due to combustion. A heater installed without its own flue (stack) will produce a *wet* heat and will cause condensation by burning up the oxygen and releasing carbon dioxide in the cabin. Carbon dioxide can cause suffocation if inhaled in large amounts. If oxygen is restricted, then carbon monoxide forms. As little as 0.00001 percent of carbon monoxide in the air may produce symptoms of poisoning, and as little as 0.002 percent breathed for 30 minutes may prove fatal. The same fuel used in a vented stove, however, will produce a *dry* heat.

Forced air, water, and electrical heating systems deliver heat from a central source and are considered remote systems. The compartments utilizing these heaters need only have the normal amount of ventilation necessary to keep the air fresh. Individual space heaters must, on the other hand, be vented from each compartment in which they are installed and, in addition to the fresh air requirements, must have the extra amount of oxygen needed for the combustion of their fuel.

Circulating water heat is the best and cleanest way to heat a vessel. It is also the most expensive to install. The piping can be of galvanized iron, but the best installation uses copper tubing. The heat source can be a separate heater, or the galley stove when fitted with heating coils, or the engine cooling water system. In most instances, the system is installed as a single-feeder return pipe and is run within the baseboard. In each location where heat is desired, a simple radiator can be made and fitted into the line by using a pair of valves to shunt the hot water to the radiator from the main feeder line when more heat is desired. The piping continues around through each compartment and returns to the heat source. A header tank is always required. When water is circulated on a continuous basis, there is little difference in temperature from the first to the last radiator; however, if the piping has a long section of unheated space to traverse, there will be some heat loss even though the pipes are insulated. It is usual to increase the size of each radiator by a small percentage to compensate for this. The hot water system is not a quick method of heating just to take off the morning or evening chill; it is more suited for continuous seasonal use, bypassing the radiators on the warmer days but never shutting the system down. It may take several hours to bring the compartments or cabins to the desired temperature; by the same token, hot water heat requires hours to cool down.

Forced air heating requires extensive ducting that not only is costly but may also complicate the joinerwork, rendering some lockers useless. This system seems to be the most practical when a single duct can be reduced in size as it passes through each compartment and is powered by a fan of the type that can be reversed for cooling in the summer. The heat is drawn from a plenum chamber that usually surrounds the firebox. A return duct is also required in cold climates, since the cabin air can be reused provided that a filter is installed prior to its being ducted into the plenum. It is obvious that for each cubic foot of air that is forced into a compartment, a cubic foot of air must be removed.

Electric heat is reserved for large power vessels that generate electricity continuously for a multitude of other purposes, such as air conditioning, cooking, and lights. In smaller vessels, electric heat is practical only at dockside.

Individual space heaters are the most common heating method on small vessels. As has been mentioned, they must be of a type that can be vented to the outside. Aside from the choice of fuel, which more or less dictates the shape, there is always a problem of siting. In vessels normally sailed in frigid waters this should be less of a problem, since the designer should start with it and arrange everything else to suit; not all designers have sailed these waters, however, nor lived in climates where heat is a constant necessity. One seldom sees a stock design or hull where this consideration is given more than a shrug.

There seems to be a misconception regarding the need for heaters as one approaches the tropics, the idea being that constant temperatures above 75° are pleasant, and heat must be avoided. Quite the contrary, heat is needed in all climates—if not to heat the occupants, at least to dry out the vessel.

Space heaters can be divided into two groups: those burning solid fuels such as wood, charcoal, coal, and cane, and those burning liquids such as oil, diesel, alcohol, and kerosene. Propane and gasoline have been used as heating fuels, but except on yachts, it is doubtful that this type of fuel will pass any maritime regulations. Even to suggest its use for producing heat results in violent tremors from insurance underwriters, although it is used for heating water and cooking.

All heaters must be constructed so they can be securely bolted to a bulkhead or secured with turnbuckles to the cabin sole. In the case of solid fuel, the fire door must open either forward or aft. The flue must have a proper smokehead that prevents downdrafts or turbulent air from affecting the flame, and it must be sited so that it will not foul any rigging nor become immersed in a normal seaway. The major axis of liquid fuel tanks must be in the fore-and-aft direction.

The solid fuels have many advantages in that a variety of fuels can be used in the same firebox without modification of the grates; the latent heat of the fuel embers makes them efficient in drying out the vessel; and numerous manufacturers provide a selection of sizes to suit almost any need. Solid-fuel stoves can be built by the vessel's builder from heavy sheet metal, and the dual-chambered European-type airtight stoves are real fuel misers. Solid-fuel heaters range from heavy cast iron, firebrick-lined, to the very light sheet metal stoves used in camping. The disadvantages of solid fuels are that, without exception, they are dirty, form creosote in the stacks, require more space per Btu than liquid fuels, and require larger stacks.

Firewood obtained from the beaches has enough salt content to accelerate corrosion in the firebox, especially the grates. Deadwood found away from the beaches makes good firewood, as do scraps and slabs from sawmills.

Inexpensive charcoal is available in many areas of the world, and it does provide good, almost smokeless heat. The carbon content ranges from about 100 percent for sugar charcoal to about 10 percent for bone charcoal. The residual ash is less than that of wood and burns slower. The charcoal briquettes sold in the United States are not only expensive but of a lesser carbon content than the charcoal of most island and South American countries.

Coal, when available, is the preferred fuel—but not just *any* coal. The best, without a doubt, is anthracite, although it is difficult to find in many areas. Anthracite coal is almost smokeless and produces little or no soot and only a small amount of ash, which is appreciated by all hands.

Coal is classified by rank based on the geological age of the deposit, as indicated by the percentage of carbon and volatile matter and the heat content in Btu's (British thermal units) per pound. Table 2 shows this classification.

TABLE 2. Classification of Coal Fuel

Rank	% Fixed Carbon	% Volatile Matter	Average Btu
Lignite (brown coal)	----	----	7,000
Sub-bituminous	----	----	10,000
High-volatile bituminous	69 or less	31 or more	12,000
Medium-volatile bituminous	69–78	22–31	*
Low-volatile bituminous	78–86	14–22	*
Anthracite	86 or more	14 or less	*

Classified by fixed carbon. The higher the carbon, the greater the Btu value.

The volatile matter of coal is composed of molecules that are converted to a gas or liquid when the coal is heated; the greater the percentage of volatile matter, the smaller the percentage of heat-producing fuel. A low grade of coal containing too much volatile matter *can* cause a stack fire and *does* cause soot accumulation, which will eventually plug up the stack. Furthermore, the volatile matter is very corrosive and will attack Dacron and nylon as well as the heater itself.

Solid fuel should be stowed in a dry, well-ventilated compartment, because all solid fuels are subject to spontaneous combustion. Coal and charcoal are best stored in metal-lined compartments, which will prevent the dust accompanying these fuels from seeping into the bilge. Newly mined coal and coal that has recently been broken up are the most dangerous in that they absorb oxygen from the air; heat is evolved during the absorption process, and this can cause the coal to ignite. Fresh charcoal is even more dangerous in this respect.

Table 3 lists other common fuels and their thermal yields. A further thought on heaters is that, with respect to conversion of the fuel to heat output, the solid fuels are about 50 percent efficient and the liquid fuels are about 65 percent efficient. The literature can be misleading in that stoves are often rated according to their performance in a high ambient temperature. This might suggest that a small stove will heat a compartment or cabin under all circumstances. What is really important is the number of Btu's required to raise the temperature of a cabin by a specified amount, and then to maintain the higher temperature. This is dependent on the materials used in the joinerwork, the insulation, air spaces, cushions, rugs, necessary ventilation, and the bodies occupying the space. The loss of heat per square foot from the heating source is proportional to the difference between its temperature and that of the surrounding air. That loss is by both convection and radiation.

Many solid-fuel and diesel cooking stoves have heating coils built into them, and thus can provide hot water both for galley use and for heating. It is also possible with these stoves to utilize small fans below the level of the stove top to aid in heating a cabin. If the galley is separated from the cabin by a full or partial bulkhead, a simple ducting system may be employed; this should be fitted with a charcoal filter to remove any grease resulting from cooking in the galley.

TABLE 3. Other Solid and Liquid Fuels Used on Vessels

Fuel	Thermal Yield
hardwoods	7,100 BTU per pound
white pines	7,200 BTU per pound
firs	7,400 BTU per pound
heating oil	140,000 BTU per gallon
alcohol	196,800 BTU per gallon
propane	208,200 BTU per gallon
butane	229,400 BTU per gallon
kerosene	275,200 BTU per gallon

REFRIGERATION

Very few vessels sail nowadays without some sort of ice chest, refrigerator, or freezer compartment. These compartments are, for the most part, not only inadequately constructed but improperly sited. The fault for this is shared by builders, owners, designers, and sellers in that they are all reluctant to deprive the interior of just one more locker, another berth, a walk-through, and so on, ad infinitum. The cubic of the correct insulation required for refrigeration often exceeds the cubic of the compartment; this is especially true when one is seeking the maximum efficiency in a minimum space.

The general practice of rating refrigeration systems on the basis of an ambient temperature of 60° Fahrenheit (taken as the mean temperature of the seawater plus the air) yields false efficiencies. In order to maintain the same efficiency when a vessel ventures into the tropics, with a mean temperature of 82°, almost 6 inches more insulation is needed. A vessel coming from up north may boast that it takes just 30 minutes a day of running the engine to maintain the refrigeration, but in Trinidad this becomes two hours in the morning and two hours at night, which is eight times that used in, say, Boston, Massachusetts. This is then compounded by using the main engine to drive the compressors for a total of 1,460 engine hours a year. It is not unusual to find the refrigerator next to the engine compartment, the stove, or both. That this may be avoided never seems to be considered.

In the tropics, 6 inches on the vertical portions of the boxes that are not in contact with the shell, 9 inches at the shell, 5 inches on the top, and 12 inches on the bottom will be found to be the minimum amount of insulation needed for either ice chests or refrigerators. Freezer compartments need considerably more. Plank urethane foam is the best insulation to use, with all joints staggered and glued to form a cavity of the size required. A vapor barrier of PVC (polyvinyl chloride) or Mylar should tightly line the cavity and be glued in place with no air entrapment. Immediately inside this is laid a reflective layer of aluminum foil with perfect adhesion to the PVC or Mylar. The inside liner of the cavity (box) can be fiberglass, metal, wood, or a combination of these. A top-opening chest or refrigerator is the most efficient type to use.

PRESERVING FISH

Fish holds are insulated with a variety of materials. At one time, cork was the primary insulating material used, followed closely by mineral wool, fiberglass, cellular glass, and charcoal. Today, sprayed-in-place urethane foam is becoming more favored, since it is easier to work and there is less possibility of voids, which are heat leaks. Once the foam has been sprayed to the desired thickness, the hold is smoothed to a uniform surface by using a heat wire and disk sander, and the whole is covered with thin marine plywood, which is then fiberglassed. Holds constructed in this manner are easy to clean and have fewer sanitary problems than do those made of other materials.

In the tropics, if the catch is chilled in brine tanks sited on deck while awaiting gutting and then thoroughly washed and chilled again prior to packing in the hold, the ice will last much longer and the bacteria count of the fish will be much lower. This method is quite attractive on sailing vessels, since the refrigeration machinery may then be deck-mounted and need not be of great capacity.

Ice and refrigeration combined offer another feasible method of maintaining the catch for longer periods of time, since the refrigeration need only maintain a temperature that will retard the melting of the ice, while the ice preserves the catch. The expense of any mechanical system must reflect the initial cost of installation, the maintenance of the machinery, and the cost of fuel to operate the plant versus the price received by the fisherman. The price of fish in the market has little bearing on the price that is paid to the fisherman, who takes all the risks at just enough profit to keep him fishing in hopes of better times.

Fish penboards should be of the removable type and may be constructed of aluminum extrusions or wood. If they are of wood, the boards should have their edges and ends chamfered or well rounded and their surfaces sanded smooth, after which two coats of a two-part polyurethane varnish are applied. Ladders into the holds should be of the rung type, with the rungs spaced not less than 10 inches nor more than 12 inches apart; the stringers should have a minimum separation of 14 inches.

The principal methods used for preserving the catch in small vessels are:

Live. This is common in the tropics, the vessels having a live well built into the hold. The method of fishing is by traps and handlining. The length of time from catching to market seldom exceeds seven days. If by chance it should, then part of the catch will be "corned" (air-dried) or salted. Some of these fishing smacks will fish and conch on the same trip. The conch are carried on deck in groups of five wired together by piercing a hole in each shell. The conch are wet down frequently during the day, and when the smack anchors each night, they are thrown overboard to keep them alive. Wiring them together keeps them from going anyplace. They can be likened to politicians—in a group they can't make up their minds which way to go.

Crushed ice. The fish may be ungutted if the trip is measured in hours; otherwise, it is customary to gut and clean them and pack them individually in ice, being certain that there is separation not only between each fish but between each layer. The holds must be fitted for drainage of the melting ice. Three weeks is about the limit for a trip.

Precooled. Crushed ice is still used, but the fish are cooled in chilled seawater so that they are near the temperature of the ice when stowed. This will extend the voyage another 10 days beyond the time it could have taken had the fish been only iced.

Refrigerated icing. This involves the addition of refrigeration piping to the hold. On motor vessels the additional power required to do this does not materially add to the cost of a voyage; on a sailing vessel, it does.

Brine freezing. This method is used by some sailing vessels and by many auxiliary sailing and motor vessels. It uses a brine of salt-fortified seawater chilled to as low as 0° F. This is then sprayed as a mist at the top of the hold, draining to a sump where it is strained and recirculated. In a 65-foot schooner of my design with a hold capacity of 45 tons, the total volume of the solution was less than 100 gallons. In most instances, the fish are precooled in a brine tank. The equipment needed to do this is small and thus very energy efficient.

Contact freezing and flash freezing. These systems are used on many small motor vessels that handline with one- to three- or four-man crews. This permits them to package the fish fillets in convenient units for institutional resale. The advantage is that the freezing process need only be used when there is a catch to be handled rather than on a continuous basis. The fish holds are refrigerated from 0° to 18° F.

The drying and smoking of fish at sea via ovens, dehydrators, or both is still in its infancy, but holds a great deal of promise. A minimum of refrigeration would be required, and the holds could have greater capacity due to lesser insulation requirements. This would prolong the time on the fishing grounds and, in all probability, would permit smaller vessels to be competitive. Undoubtedly, some advancements will be made in irradiation and other methods of preservation that will be compact and not lethal to either the crew or the consumer.

2

▽

JOINERWORK

A builder is judged as much if not more by the quality of his joinerwork than by his ability as a steelworker. Joinery may be defined as cabinetwork that does not add nor seek to add structural strength to a vessel. It may be either internal or external woodwork. A joiner in the days of wooden shipbuilding was not a ship carpenter; he was involved only in work with light materials that needed a fine finish and appearance when the work was completed. When not employed in shipyards, many joiners plied their skills in other industries, such as making antique reproductions, furniture, and cabinets.

Some definitions must be modified when they relate to steel boatbuilding, for it is generally difficult to determine the line of demarcation between one job and another. Decks and deckhouses constructed of wood in a steel vessel do not add appreciable strength to the vessel and are therefore considered joinerwork. Structural bulkheads of any material other than steel are not considered a strengthening portion of the hull; a watertight wooden bulkhead that substitutes for a steel bulkhead, totally dividing the vessel on a frame and forming a major compartment on either side, will be considered a structural bulkhead, hence joinerwork. Laminated plywood cabintops covered with fiberglass are considered structural portions of the hull and therefore are *not* joinerwork. Cargo hold ceiling and sparring are joinerwork, but wooden cargo hatch covers are considered carpentry work.

In commercial vessels, the exterior joinerwork may be nonexistent in that everything that would normally have been constructed of wood is built of metal. This reduces maintenance, since there is only one material with which to contend. Yachts, on the other hand, may have wooden decks, cabin trunks, coamings, rails, seats, bitts, cleats, hatches, skylights, doors, grabrails, ladders, companionways, belaying pins, and so forth. The cruising vessel usually falls somewhere between these two extremes.

INTERIOR ARRANGEMENT PLANS AND JOINERY

Stock plans include at least one and sometimes several arrangement plans, which are furnished by the designer as guidance plans for the builder and owner. It is always assumed that any and all modifications or changes from these plans will be made by the builder or owner, and not by the designer unless he is commissioned to do so. When designing the original interior arrangement of a particular vessel—which usually becomes a stock design when several owners cover the designing costs jointly—the designer incorporates as much as possible of the owner's original requirements, while rejecting other ideas as impractical or impossible. At that time the design is fresh in the mind's eye of the designer, and he can easily visualize the cubic space distribution; 10 or 20 years later, however, few can recollect the reasons for decisions to incorporate or disregard particular features of an interior design. Interior arrangement plans consume a major portion of the time to prepare a complete set of plans, since they must be in agreement with all other plans.

In steel designs, where each frame is drawn on the construction sections plan, it is now customary to show some or all of the joinerwork on the same drawing. How detailed the joiner plans need to be depends mostly on the type of joinery used. The modern "public washroom" motif needs very little in the way of details, while the "rococo" motif might require that every piece of joinery be fully detailed.

In order to prepare a proper arrangement, the designer needs not only personal experience at sea but often an intimate knowledge of the owner's habits. It is rare when an arrangement prepared for one vessel can be lifted *in toto* and used on a completely different design, for each vessel has its own inherent peculiarities. It might be charitable to say that studying the arrangements of production vessels can provide a good guide for an owner and builder, but it does not, for production vessels are intended to appeal not to the individual but to the mass market. One must also realize that the majority of people suffer in some degree from claustrophobia, and this will affect the amount of joinerwork that is possible in any given design. The test of any arrangement is how many changes must be made, say, after one or two years of living with and in it. It is not a rare occurrence for the owner to make minor and sometimes even major changes to the interior after living with the original arrangement. This in no way reflects on the abilities of the designer and the builder, as they can only advise the owner as to the practicability of his desires. Neither can the blame be attributed to the inept thinking of the owner, for, when one is living on land, there is a tendency to mix up priorities and apply them in the wrong order to a vessel for use at sea. The ideal time to design an interior is when one is at sea, because then the order of priorities for the whole interior becomes acute, and any inconvenience is magnified beyond the importance it assumes when one is in port.

Even custom design arrangements must not be followed blindly, as there is seldom an instance where slight adjustments will not improve the overall efficiency and comfort of a cabin. No graphic geometry on a drawing board can equal the actual occupation of space for making one aware of potentials or limitations.

The only caution in making changes to a stock plan is that major weights, such as the engine or tanks, must not be moved without the designer's approval; otherwise, one is free to do just about anything that is imaginable. There is a finite length, breadth, and depth available in any vessel, which prevents one from doing much harm to its trim.

When designing and constructing a vessel's interior, the basic fault made by many is in the abrupt discontinuities of major joiner items caused by accenting some of the details. The eye should encompass the whole as a blended unit, and then have to focus to discern a detail. For the professional builder, some mock-up work is necessary in most vessels to assure that there will be no grossness in the interior. The amateur builder may have to be a bit more elaborate in the amount of detailing that needs to be mocked up, but it is better to do this than to have to remove a piece of joinerwork at a later date.

MODIFICATIONS TO INTERIOR ARRANGEMENTS

The Pinky Schooner used to illustrate the building of a hull and decks in Volume 1 was chosen partly because of my fondness for the type, but mostly because the steelwork used in the construction of her hull form covered the usual as well as unusual problems that occur in steel construction. Several people have built this Pinky, so her Interior Arrangement Plan (Figure 1) will be used to indicate some of the changes from the original that were required by other owners. The changes illustrated have by no means exhausted the possibilities, because each owner may have a different priority for the available space.

The planning of a different arrangement for a particular owner is not difficult, since decisions have already been made as to the hull design and the rig. Suppositions of "if it were longer," "if it were wider," or "if it were deeper" have no bearing, since they imply an entirely different vessel. One may liken this to sailing a vessel that is engineless. In any situation the question is never "What would I do if I had an engine?" There is no engine, and the relevant question is what must be done with what exists.

Before planning specific changes in an arrangement, the builder must consider that the deck arrangement has a direct bearing on what can be done below decks. Widening the cabin trunks in the Pinky serves no useful purpose and in fact would be objectionable, because the reduction of deck area would be a loss of deck cargo space and would make the vessel difficult to work. In all vessels, and in sailing vessels in particular, the ability to move freely fore and aft without worrying about one's foothold is more important than a few extra inches of room below decks in an area where one cannot stand.

The collision bulkhead at Frame 3 of the Pinky could be moved forward approximately 18 inches and still conform to my requirement that it be 10 percent of the designed (or datum) waterline aft of the face of the stem. (Most regulatory bodies require a minimum of 5 percent, but the Pinky's stem is raked enough to give her a generous safety margin in case she should suffer a head-on underwater collision. At the 5 percent minimum, it would be possible to sustain damage at the collision bulkhead and flood both the forepeak and the forward cabin.) However, shortening of the compartment forward of the collision bulkhead can be done only

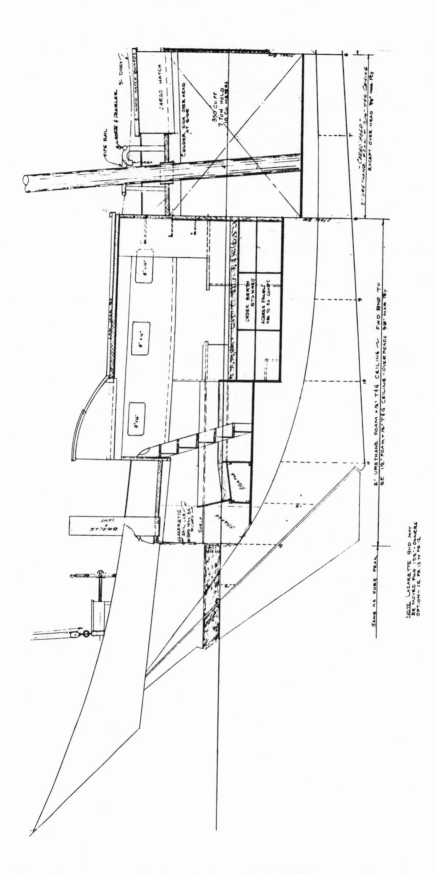

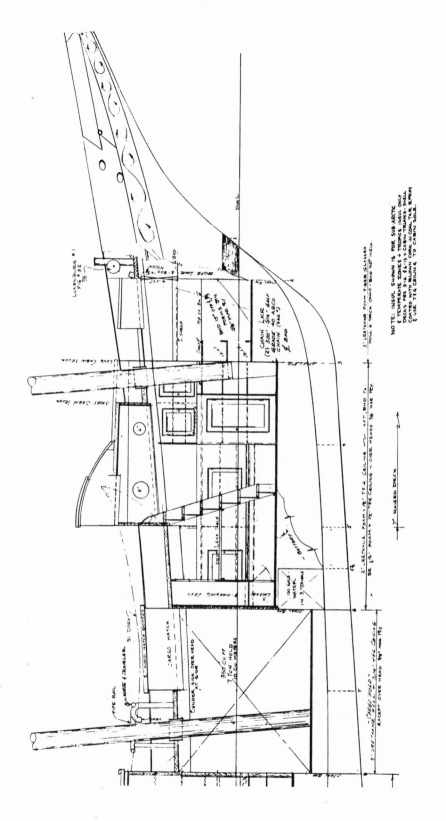

Figure 1. *Reproduced on these four pages is the Interior Arrangement Plan of Colvin Design No. 169, a Pinky Schooner. The building of this vessel is described in Volumes 1 and 2 of this book. Length over the rails, 50 feet 8½ inches; length on deck, 42 feet 8 inches; molded beam, 12 feet; draft, 5 feet 9 inches; displacement, 38,000 pounds; hold capacity, 7 tons or 350 cubic feet.*

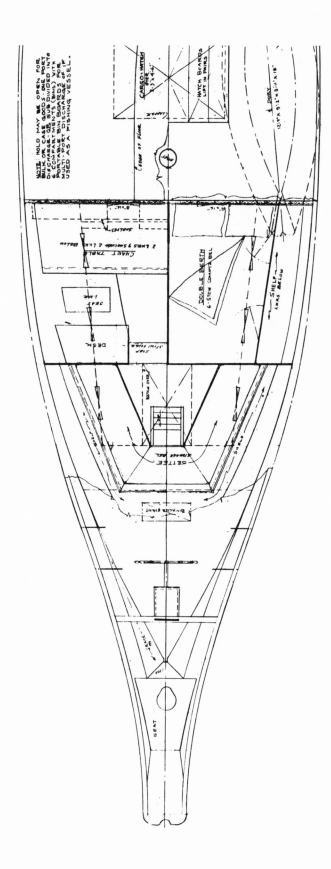

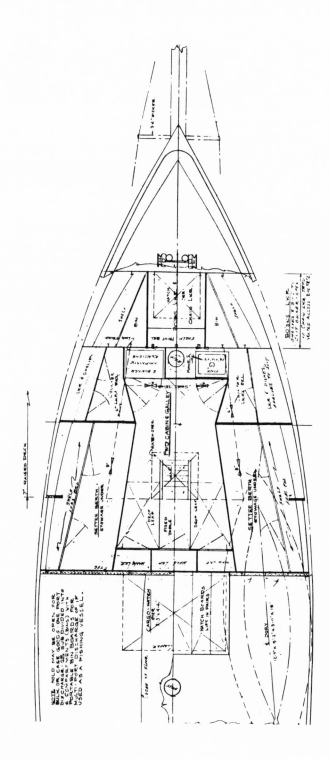

with a loss of space in the chain locker and a great loss of space for bosun's stores. If only one anchor chain is required, the remaining chain locker will suffice without raising the sole and losing what little headroom exists in this space. Many of the bosun's stores could be relocated into the bilges, but this reduces the bilge space for other items and could be inconvenient. Should such an alteration be contemplated, it would have to be made during construction.

Unless double sheeting of the jib is acceptable, the cabin trunk should not extend forward of Frame 3, since the horse needs to be longer than the forward end of the cabin is wide. If double sheeting is acceptable, the cabin trunk could extend to and join the forepeak hatch (which can be 16 inches forward of Frame 3) without necessitating other alterations to the deck structure.

The space between the forward and after cabin trunks is about the minimum for carrying a proper working boat, keeping in mind that, as designed, this vessel has no auxiliary power. About the smallest acceptable boat would be 10 feet long and 4 feet over the gunwales; therefore, about the only space available to carry it is on deck. No matter how the cabin trunks are lengthened, there will never be enough room for the boat to lay fore and aft atop either trunk without creating other problems. Carrying it athwartship is possible, but it would block the entrance into the forward cabin, create an obstacle on deck (there would be but 12 inches in the clear on either side of the vessel), cause fouling of the fore sheet, and partially block the hold ventilation. An inflatable is not an acceptable substitute.

At 6 feet 6 inches long, the hold is as short as it can be and still have any chance at all of carrying a paying cargo—that is, about 7 long tons. At best, a vessel of this length and capacity is in all probability better suited for use as a trader or fishing vessel than as a third-party freighter.

The aft cabin is as long as is practical. Going farther forward above the deck is possible but would require a relocation of the bilge pump, a modification of the hold vents, and more thinking about the working boat. To extend the cabin farther aft would necessitate narrowing the trunk sides and moving the binnacle and steering gear aft; even at that, one can extend the cabin only a few inches aft while still leaving a place from which to steer. It seems that about the only change that may be possible without sacrificing other needed space is the relocating of the collision bulkhead.

The Interior Arrangement Plan (Figure 1) indicates the original owner's preference for use in the tropics and his desire to carry cargo when the occasion arose. The aft cabin assures total privacy for the owner should he have either guests or a couple of passengers sleeping forward. The only toilet facility is the seat of ease at the stern. The galley is small, and the forward cabin trunk was extended to Frame 3 to permit the cook to bend over the sink without hitting the trunk. The large forepeak was needed to accommodate rather ample bosun's stores and some specialized working gear. The stiffening frames of the cargo hold bulkheads face into the cabins and are insulated as well as sheathed with T & G (tongue-and-groove) material. The steel bulkhead at Frame 3 is painted. The dining table is of dropleaf construction, the center portion being permanently attached to the centerline hanging locker and the companionway ladder. The vessel's anticipated days at sea are only 100 per year, but her voyages will last from three to five years. This excuses the double berth being sited in the open, where it will be well ventilated.

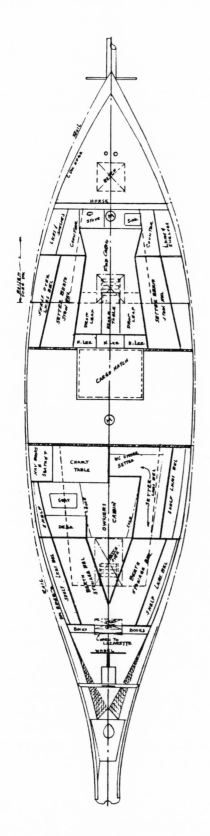

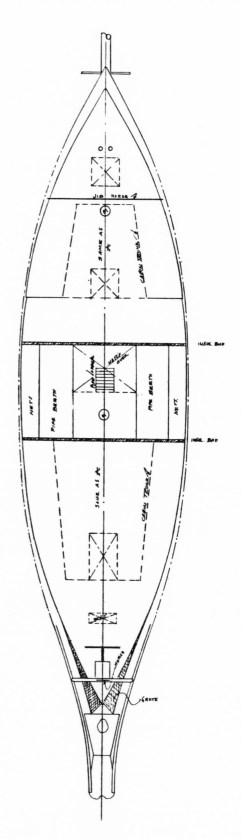

Above: Figure 2. *Plan view of an alternative arrangement for the Pinky, showing modifications to the after cabin.* **Below: Figure 3.** *A second alternative Pinky arrangement plan, differing from Figure 2 only in a modification to the cargo hatch and the addition of pipe berths in the cargo hold.*

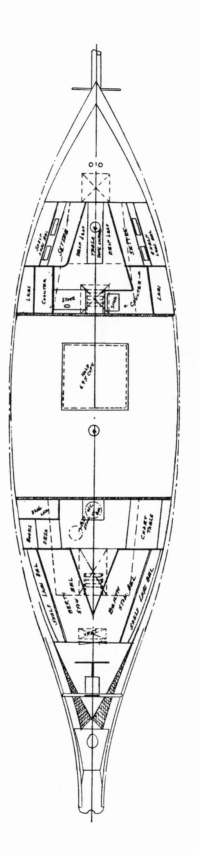

Figure 4. *Another arrangement plan for the Pinky, this one offering maximum cargo capacity.*

Figures 2–4 show some of the alternatives that are possible to a stock plan in order to suit the requirements of different owners. The plan view in Figure 2 retains all the features of the original arrangement except that the after cabin is rearranged to have two single berths; the large chart table, the desk, and the settee are retained, and a water closet has been added under the forward portion of the settee.

The plan view in Figure 3 duplicates Figure 2 except that the cargo hatch has been modified by the addition of a lift companionway hatch and an iron ladder. The hold can now be used as a stateroom on those occasions when extra guests or passengers are carried. The use of pipe berths poses no problem in that they may be dismantled when not needed. Any fixed joiner lockers or berths would need to be ripped out when cargo is carried; therefore, hammock nettings are used for clothes stowage.

The plan view in Figure 4 offers maximum cargo capacity. This is achieved by respacing the bulkheads forward and aft by an additional frame space. The galley is relocated to the forward cargo hold bulkhead frame 5, and bulkhead frame 3 is relocated 16 inches forward to gain the needed seating space. The loss of forepeak space is acceptable because the vessel seldom anchors to all chain, and the loss of bosun's stores is unimportant, since she frequently returns to areas where ships' supplies are reasonably priced.

Of all the plans, the original aft cabin arrangement has the maximum usable room, comfort, and stowage space. Assuming there is always someone on watch while at sea, which is usual, then a double berth is quite comfortable. Continual bad weather is the exception, not the rule.

When judged by yachting standards, all of the above arrangements are sparse and maybe somewhat primitive. The emphasis has been placed on light and ventilation, comfortable sleeping berths, an area in which to lounge, an area in which the crew and guests can sit down and eat, a galley that is convenient enough for preparing meals, locker and shelf space in each cabin for its occupants, a proper chart table, and a desk. The owners of commercial vessels usually have the attitude that one should always have good sitting headroom, but if there is no use for full headroom it is not necessary; one can always find full headroom on deck. Yachts, on the other hand, seem to place emphasis on larger heads and showers, maximum berthing, maximum cabin width and length, standing headroom even over berths, extra-large galleys, a minimum engine room, and a minimum of usable deck space.

INTERIOR JOINERWORK MATERIALS AND FINISH

The use of tongue-and-groove material for joinerwork involves the least amount of time, wastes the least amount of material, and gives the builder an infinite choice of grain configurations and color blends. The material should be center-matched T & G with identical beadings, if any, on each face so that it may be end-for-ended whenever desired. T & G used for ceiling is ¼ to ½ inch thick, square-edged. When used for bunk fronts, cabinets, and other short joinery, it is ½ to ¾ inch finished thickness, and the edges are veed. When used for bulkheads, ¾ to 1 inch finished thickness is required. All the tongue and groove required for the vessel should be from the same mill run or, if milled by the builder, should be done at the same time, to avoid the possibility of a mismatch in the tongue, the groove, or both.

Occasionally an owner will desire the use of random-width rather than uniform-width material. This creates a time-consuming task for the builder, for each bulkhead must be figured separately. For example, if a bulkhead of 107 inches (8 feet 11 inches) were required, with an allowance of ½ inch at each end for the removal of the T & G and trimming, the usable faces of the material might be 3½ inches (1), 5½ inches (2), and 7½ inches (3). The bulkhead would then have the following pattern: 3, 1, 2, 1, 2, 3, 2, 1, 2, 3, 2, 1, 2, 3, 2, 1, 2, 3, 1, and 2. A more common pattern is 1, 2, 3, 1, 1, 2, 3, 2, 1, 2, 3, and 3; however, this is not random, and in long bulkheads the pattern becomes so repetitious that it may be associated with grooved plywood or contact paper.

The interior finish is a matter of personal choice, but several things are worth noting. A natural finish that is varnished requires the least maintenance and will, if oiled down once or twice a year, last at least 20 years without revarnishing. To obtain this longevity, one must start by applying several coats of tung oil to both faces of the wood; then the portion that faces the cabin must receive a minimum of four and preferably eight coats of varnish that has an oil base compatible with the tung oil. Each coat should be wet-sanded prior to application of the next. Any attempt to use a surface sealer and synthetic varnishes is inviting trouble because, when moisture penetrates the wood, the finish will lift off in sheets. Painted surfaces, too, will benefit by using tung oil as the sealer, and then building up with a flat coat of paint prior to applying the gloss coats. Linseed oil (boiled) has a tendency to mildew, while tung oil does not. Other penetrating oil sealers on the market have modifiers added that make them acceptable substitutes for the pure oils.

In the tropics, a vessel's interior may and should be darker than the same interior if used in a temperate zone. The closer one is to the equator, the more intense the sunlight and reflective glare become. When the overhead is painted white or some other light color and there are enough portlights for good ventilation, there is more than enough light to read by during the day without artificial lighting, even when the darkest woods are used for the joinery. Some vessels sailing in the tropics are so bright below that one needs to wear sunglasses to be comfortable.

Decks should be painted white, since even the slightest tint will absorb heat, making the vessel uncomfortable not only below but on deck. Glare is never from the deck, but from the water and the sun. At night, white decks are easy to work on without additional lighting.

There is no such thing as ugly wood, but man in his infinite wisdom can make it ugly. Staining one wood to look like another will always make it look like one wood stained to look like another. Wood stained or bleached to have an even appearance always looks artificial. The proper way to achieve different tones and textures is either to use different species of wood or to select boards that are lighter, darker, or of different grain from the same species. In selecting various species of wood that are to be combined and finished natural, one should seek a harmony of colors that will be enhanced by the proper selection of upholstery, linens, tiles, rugs, and curtains. Woods with a reddish finish (ash, basswood, birch, cherry, elm, honeylocust, maple, red cedar, redwood) are accentuated by white, black, and yellow; those having a bluish and purplish tint (magnolia, black walnut, yellow poplar, pine) are accentuated by white or yellow; those having a greenish tint (black locust, magnolia, yellow poplar) are accentuated by black, white, and yellow; while those that have a gold color or flecks of gold and different shades of light brown (aspen, birch, black locust, cypress, cedar, pine) are accentuated by white, black, brown, blue, purple, and pink. The more popular woods for interior joinery are teak, Honduras mahogany, rosewood, silver bali, butternut, oak, African and Philippine mahoganies, black walnut, cherry, and yellow cypress. Often other local woods, such as the pines and cedars, are suitable for interior joinery; however, the best woods to use for the exterior are teak, silver bali, delmari, Honduras mahogany, dense Douglas fir, and longleaf yellow pine.

INTERIOR JOINERWORK DETAILS

Structural Bulkheads

In many of the smaller vessels, steel bulkheads are impractical due to their weight and cost of construction if they must also be sheathed with wood in way of the accommodations. Watertight wooden bulkheads will then be substituted, using a normal steel frame. In this construction, all longitudinals pass through the frame using the same detail they would use if passing through a bulkhead. The frame is welded on both sides to the shell. The construction of a wooden bulkhead to replace one of steel is done in three phases. First, a ¾-inch marine plywood bulkhead is made so that it laps the frame and rests against all longitudinals. A piece of wood is then cut for the opposite side of the frame, the wood being wide enough to counterbore for a

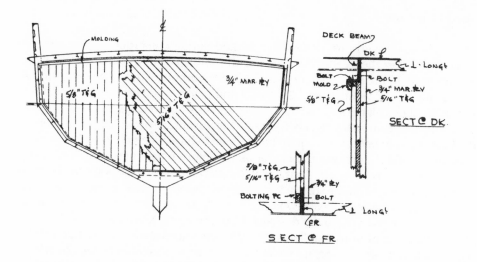

Figure 5. *Details of a ¾-inch marine plywood bulkhead overlaid by a diagonal tongue-and-groove ceiling and a second, finish layer of tongue and groove.*

nut and washer, and bolting holes are drilled through the bolting piece, frame, and plywood at 6- to 8-inch intervals. After drilling, all wood is moved clear of the frame, any steel chips from drilling are cleaned off, and the faying surface is well luted. The bulkhead is then bolted to the frame using galvanized carriage bolts with the heads on the plywood side. In the second phase, a T & G ceiling having the same thickness as the frame is laid diagonally across and glued to the plywood. Third, this diagonal layer is overlaid with the finished wood that will be seen in the cabin. This material must also lap the frame and will butt on the bolting piece; it will be luted where it fays a steel surface, and glued elsewhere. Figure 5 details this type of bulkhead.

Watertight subdivision to prevent sinking in case of holing is virtually impossible in small vessels with their accommodations below the main deck, because the required compartment lengths are too small to be usable. At best, the subdivision will confine the water to a single compartment, making it easier to pump. The fallacy of single-compartment subdivision is in the supposition that holing will occur *only* in one compartment, which is a fine notion when one is ashore sitting at a desk. In actual practice it is best to suppose that the worst will happen and the damage will occur at the bulkhead, causing the vessel to sink. In large vessels, 600 feet or longer, the single-compartment subdivision has more validity. The strength of a small steel vessel can only be likened to that of a battleship; it can tolerate strandings, collisions, and other mishaps that would destroy vessels of other materials built to the same design and would severely damage or destroy a large steel vessel. The primary purpose of watertight bulkheads in small vessels, then, is to isolate the various compartments from each other, preventing odors and contaminants from seeping through—for example, to the forepeak from the forward cabin, to the cargo hold from a cabin, or to the engine room from a hold or a cabin.

The Cabin Sole

The cabin sole beams are sometimes omitted during construction of a vessel to allow the builder the freedom of adjusting the sole up or down, thus gaining either greater width to the arrangement or increased headroom. If this is the case, the sole beams must be of wood, since all blasting and painting will have been done when they are installed. If the span is short, 2 x 4s are used; for longer spans, 2 x 6s, 2 x 8s, 2 x 10s, or 2 x 12s are used (Figure 6). It is to the advantage of the builder to keep the depth of the sole beams at a minimum in order to have as much clear space in the bilge as possible; therefore, the use of one or more pillars per beam is in order. Wooden beams are bolted to the frame using galvanized carriage bolts, the heads of which are best sited on the steel side so that the paint will not be gouged or chipped when the nut is turned; at least two bolts per side are required. In some vessels the cabin sole is not parallel to the designed waterline (DWL), in which case the beam tops are beveled prior to installation. All wooden beams should be painted prior to final installation, and the faying surface against all steel should be well luted.

If sole beams of steel (usually angles) were installed at the time of construction, there are three options for the builder. First, he may drill a bolt hole in each beam for each piece of sole material. This will cause trouble in the long term, since the cabin

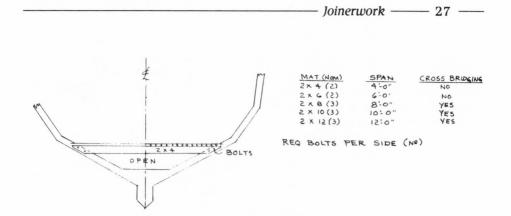

MAT. (NOM)	SPAN	CROSS BRIDGING
2 × 4 (2)	4'-0"	NO
2 × 6 (2)	6'-0"	NO
2 × 8 (3)	8'-0"	YES
2 × 10 (3)	10'-0"	YES
2 × 12 (3)	12'-0"	YES

REQ BOLTS PER SIDE (N⁰)

Above: Figure 6. *Required sizes and details of attachment for wooden cabin sole beams of various spans. Two bolts are required at each end to tie 2 x 4s and 2 x 6s to steel frames; three bolts are required for larger beams.* **Below: Figure 7.** *Three methods of attaching a cabin sole to steel sole beams. Either of the bottom two methods is preferable.*

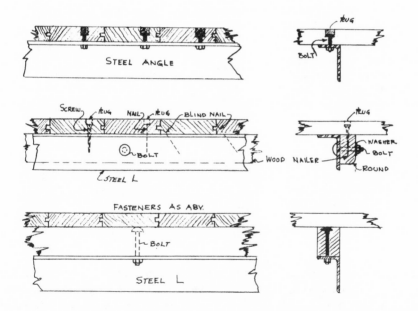

sole should not be laid in luting. The second method is to bolt to the side of each angle a wooden beam, and then fasten the sole to these. Third, if headroom permits, wooden fastening pieces may be bolted to the tops of the angles. In any case, the wood that accepts the cabin sole fastenings must be at least twice as deep as the cabin sole is thick. Either of the latter two methods or substitution of the 2 x wooden beams for the steel beams is the preferred type of construction, since once the wood is laid the cabin sole is all wood-to-wood joinery, and no further precautions need be taken to protect the steel. Figure 7 shows these details.

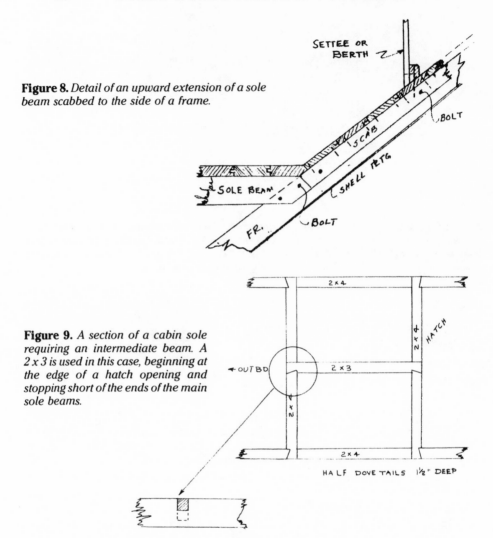

Figure 8. *Detail of an upward extension of a sole beam scabbed to the side of a frame.*

Figure 9. *A section of a cabin sole requiring an intermediate beam. A 2 x 3 is used in this case, beginning at the edge of a hatch opening and stopping short of the ends of the main sole beams.*

In smaller vessels, in order to keep the freeboard within reason, it is necessary for a portion of the cabin sole to climb the frame to meet the longitudinal joinery, especially at the forward and after ends of the vessel. When this occurs, a wooden piece is scabbed to the side of the frame as a continuation of the sole beam; Figure 8 shows this detail. These scabs do not have a constant bevel, since the side sole when laid is parallel to the shell and not the DWL. The sole that is laid on the scabs need not be as thick as the main sole, but it should be thick enough to support a person when the vessel is heeling.

When the transverse frame spacing exceeds 24 inches and the cabin sole is less than ¾ inches undressed thickness (1 inch finish thickness), there will be a need for an intermediate beam or beams to prevent any springiness in the sole as laid. Figure 9 shows a section of a cabin sole where this is required. These beams need not be as heavy as the beams that are bolted to the frames, nor must they extend all the way to the shell; rather, they may stop at or just outboard of the area upon which people may walk. Outboard of this a cleat is usually located, to which a transverse partition or cabinet is attached. This will stiffen the sole enough to accept any load that could be located there. Toward the centerline, a header should be placed wherever there is a hatch in the sole.

The best practice is to lay the cabin sole from one structural bulkhead to the next as one continuous flat that will support all the joinerwork. If sometime in the future a major alteration to the cabin arrangement should be required, then a new sole would not have to be laid. A practice among some builders is to let the cabin sole be intercostal to (interrupted by) the joiner bulkheads, and to use plywood as the cabin sole. In this way thin plywood bulkheads are stiffened, and the baseboard nailing strip that is necessary with a continuous sole is eliminated. When major rebuilding is undertaken, however, a complete new sole is often required.

If the vessel has been properly built, painted, and insulated, there should be no reason to take up the sole for either inspection or maintenance of the hull, provided there are sufficient hatches in the sole to permit one to see and reach all areas that are covered. The minimum width for these hatches is 10 inches in the smallest vessel with a narrow sole to 20 inches in the largest vessel with a wide sole. In the larger vessels there may be enough depth below the sole for a workman to enter the area. Opening the entire length of the sole in one or several pieces is seldom possible or convenient, but as long as the maximum longitudinal area that cannot be opened never exceeds 24 inches, one should experience little difficulty with access to the bilge area. If this maximum has to be exceeded, then it is necessary to have a hatch off the centerline to remedy this.

A thick cabin sole has some advantages in that it needs less framing, adds weight down low, and makes possible the use of double T & G along the planks. (Figure 10 shows this standard mill joint.) Plank cabin soles are made of Douglas fir, Honduras mahogany, black walnut, cherry, silver bali, or teak. These planks may be laid edge to edge with T & G, square, or separated by a thin strip of another material of contrasting color, such as holly between teak planks or mahogany between fir planks. The separation strips should be between $^{3}/_{16}$ and $^{3}/_{8}$ inch thick; otherwise, they will detract from the remainder of the sole. Plywood can also be used for the cabin sole, with teak veneer, teak or oak parquet, thin teak planks separated by narrow holly strips, linoleum, rubber tiles, or vinyl tiles glued to its upper face. The use of carpet over plywood is increasing on yachts. This produces some saving in labor for the builder, and one must suppose that this is better than laying a carpet over a finely finished sole.

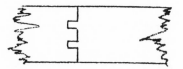

Figure 10. *The standard mill joint used for double tongue-and-groove planking.*

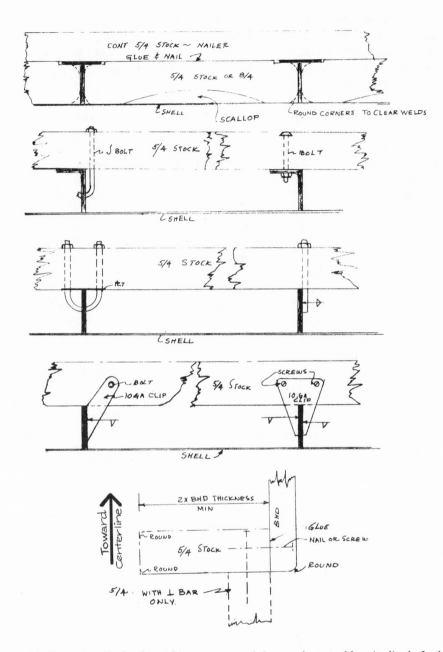

Figure 11. *Several methods of attaching transverse joinerwork to steel longitudinals. In the top four sketches, one is looking forward or aft at a sectional view through two longitudinals on the bottom or (if the sketches are rotated 90 degrees) the side of a vessel. The scallop cut out of the furring piece in the topmost sketch permits drainage. In the bottom sketch the view is from overhead, with the vessel's centerline off the top of the drawing. The furring piece shown in cross section at the bottom of the drawing would be used with T-bar longitudinals only. The width of the nailing strip must be at least twice the bulkhead thickness.*

Joiner Bulkheads

Joiner bulkheads are constructed after the sole is laid. It is usual to begin with the transverse ones, and then connect the longitudinal bulkheads to them as required. Transverse plywood bulkheads are attached directly to the frame with several bolts; as always, the faying surface to the steel is luted. When T & G material is used, it is glued and fastened to a nailer strip that is bolted to the frame, thus eliminating the numerous bolts that would otherwise have been needed. Small partitions will seldom fall on a frame; indeed, joiner bulkheads commonly fall well clear of any frame. If the vessel is longitudinally framed with Tees, the solution is a simple one that involves neither bolting, welding, nor drilling into any frame. It requires only a furring piece with a length equal to the distance between longitudinals and a depth equal to that of the longitudinals; the upper edge of the piece is notched at each end to slip under the flanges of the longitudinals. A continuous nailing piece is then fastened over the tops of the furring pieces, longitudinals, *et al.* When angles or flat bars are used for longitudinals, either a clip must be welded to the longitudinal or a hole must be drilled, and a J- or U-bolt is used. Figure 11 shows several methods of affixing transverse joinerwork to metal, and is self-explanatory. Welding bolts to frames is a poor method unless special bolts are used, since the standard bolt from a supplier has a high sulfur content and becomes brittle when welded.

All subsequent joinerwork fastens to the joiner bulkheads, partitions, or shelving. This will further stiffen the bulkheads at various levels depending on where a berth bottom, shelf, or partition is attached. The interior of a vessel is akin to a jigsaw puzzle in that each piece by itself is strong; partially assembled the structure is weak; and when finished it is once again strong.

All cleats, nailing strips, shelf and bunk supports, and indeed all edges of wood that are not covered with a molding should be chamfered or rounded, since this lessens the likelihood of splinters and cuts.

Ventilating the Joinerwork

All lockers and compartments must have ventilation of some sort; otherwise, the area is a haven for mildew and other fungi and provides the necessary conditions for rot. A well-ventilated area will be perfectly safe for clothing, canned goods, books, papers, blankets, and so on. This is accomplished by using ventilated doors, vents drilled into settee and berth fronts (Figure 12), a space between the shell and all shelving, or grates, louvers, caning, basketweave, and other such constructions to allow free circulation of air. Cutouts are frequently used, but this becomes complicated unless one is content to use anchors, fish, and other nautical objects, which seem more suited to the den, bar, or kitchen ashore to suggest that one is in a vessel. At sea there is never any doubt, so one does not really need this illusion.

One of the most economical systems of ventilation is the use of basket-woven doors, since the material can be cut from scraps of joiner material that have no other possible use in the vessel. The strips are cut approximately $\frac{1}{32}$ inch thick, with their width being the same as the thickness of the joiner material. When a circular saw with

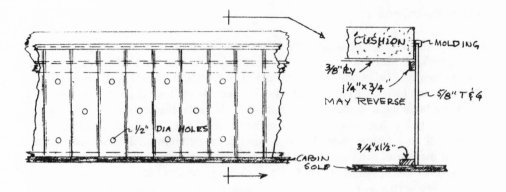

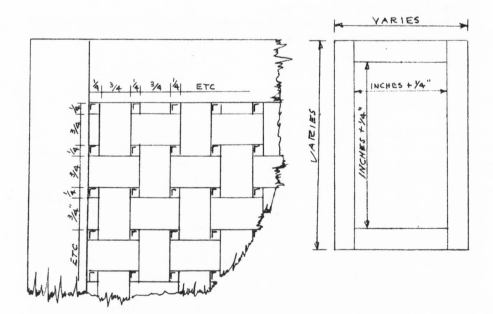

Figure 12. *Half-inch holes drilled in a regular pattern in a berth front provide ventilation. Other berth details and recommended cushion thicknesses are also shown.*

CUSHIONS

SETTEE	3" MIN	4" MAX
BERTH	4" MIN	5" MAX
SEAT	1½" MIN	2" MAX
DECK	2" MIN	4" MAX
BACK	1" MIN	2" MAX

Figure 13. *Detail of a basket-woven door.*

carbide teeth (72 teeth on a 10-inch blade) is used, the resulting strips can be woven directly without sanding. The secret to using these strips is to start with a space and end with a space. The spacing is usually ¼ inch, which is close enough to assure privacy between one compartment or the head and another compartment. The strips are stapled to the back side of the door using stainless steel or bronze staples of the size used to staple paper. Over this, a strip of wood is nailed all around, which not only keeps the strip ends from snagging but has a tendency to tighten them as well. Gluing is not necessary when basketweave is used, and the varnish soaks right through the strips, making them even more rigid. Figure 13 details this type of door. The overall length and width of the door framing will vary, but the inside dimensions must always be an even inch plus ¼—for example, 9 inches plus ¼ inch, or 16 inches plus ¼ inch—when ¾-inch strips are used with a ¼-inch spacing between them. If another spacing or strip width is desired, the inside dimensions of the door frame must be adjusted accordingly.

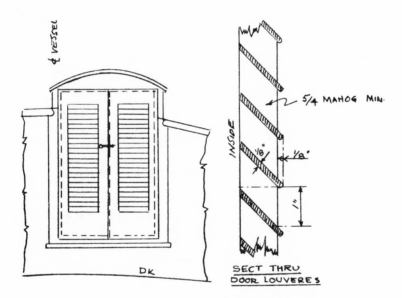

Figure 14. *The use of louvers in a companionway door.*

Louvers provide good ventilation, but when there are numerous locker doors they seem to dominate the interior rather than blend in. On the other hand, their use in companionway doors provides the best weatherproof ventilation that can be devised. Figure 14 details this type of door. The inside of a louver leading to the weather deck should be screened and, for the colder climates, should have a closing panel that fits over the screen or is used in lieu of it. Instead of wood, louvers can be made of sheet Monel, which looks fine when weathered. Polished brass is too much work to maintain, and stainless steel is difficult to work.

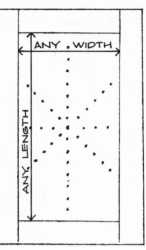

Figure 15. *One-quarter-inch plywood used as a door panel and ventilated with a decorative pattern of holes.*

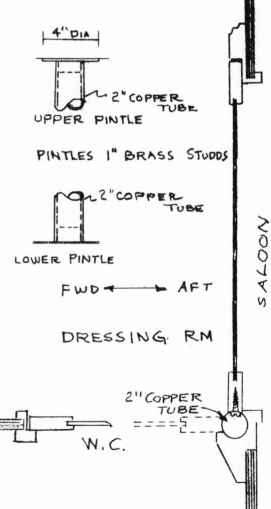

Figure 16. *A view from overhead of a single door used to separate a dressing room from either a water closet or a saloon, depending on which way the door is swung.*

Pegboard with a pebble texture also provides good ventilation; however, it varnishes almost black and is therefore best suited for use with the darker woods. A substitute for this is to use ¼-inch plywood as the door panel, and drill in a decorative design using ⅛-inch diameter holes (see Figure 15). Since the plywood panel insert may be of any length and width, the door frames in a vessel may be of a constant width.

In some instances it is possible to use one door in place of two. This can be done as shown in Figure 16, where the same door separates the water closet from the dressing room when swung in one direction, and the dressing room from the saloon when swung in the other. The openings need not be the same, for it is quite acceptable for a door to lap a bulkhead provided the trim creates the illusion that the door is an exact fit.

Fastenings

The joinerwork may be fastened in numerous ways. My standard was to counterbore, glue, nail, and bung (plug) for each fastening that was exposed to the interior, and to glue and nail elsewhere, since anything short of permanence was never intended. The use of galvanized nails that are bunged is quite acceptable, but to leave the heads exposed as in house construction would only cause trouble after a few years. There is an exception to this, and that is in the use of galvanized brads to secure mortised corners and light moldings while the glue is drying. The brads are driven partway, the head is nipped off, and the remainder is driven flush. When varnished, there never seems to be either discoloration or rust. The use of brass escutcheon pins in lieu of galvanized brads is possible, but these always show up as bright spots and do not hold as well. Some builders prefer to use screws but no glue when assembling their joinerwork, on the assumption that a totally removable interior is desirable. The screwheads then show, and some builders even use finishing washers with them.

Water Closet

The water closet sole can remain a natural finish if a shower is not installed. Showers are best sited as portable units on deck, or *anywhere* but below. Yacht practice in the United States, however, is to install showers below decks, which poses some serious problems not only in plumbing but also in ventilation. The area is always damp, and a water closet–shower combination can seldom have linen lockers or any other storage lockers. The sole must be covered with a pan of metal, cemented with a tile overlay, or fiberglassed. The shower stall should have at least a 6-inch sill, and if the stall is not a separate enclosure, the entire water closet must be made watertight to at least 6 inches above its sole.

Countertops

Countertops nowadays are usually of Formica, but in the past they were usually of hardwoods such as maple or ash. The wooden counters were kept well scrubbed, since any negligence in this task could result in salmonella, which causes sickness and, in acute cases, even death. The use of metal counters was and is quite widespread in large commercial vessels, and these counters are easier to keep clean than those made of wood. One of the most satisfactory countertops uses ceramic tiles laid in a waterproof cement with a waterproof grout for finishing. It is easy to keep clean and provides a surface that one can use for any hot pan or plate. Other satisfactory materials are battleship linoleum, vinyl tile, rubber tile, and fiberglass.

No working surface is absolutely nonskid in a seaway. Wet dish cloths or towels are frequently used to provide a nonskid surface in the galley and on the saloon table. All fiddles should be at least 1¼ inches in height and should present a vertical face on the table or counter side. The ones that are tapered toward a table surface cause a dish or plate to flip over in a seaway. Portable fiddles are sometimes used, but stowing them is a nuisance. In the long run, permanent fiddles are better.

Cargo Hold Ceiling

The cargo hold ceiling is usually laid as planks for the width of the floors, with narrower strakes up to the turn of the bilge. Above this, regular cargo battens are used. When only very light cargoes are to be carried, the use of ¾-inch T & G is permissible if it can be supported on frames or sleepers spaced 15 inches or less. In the smaller vessels, it is customary to use standard 2 x material, which nowadays means a finished thickness of 1½ inches. In the largest vessels—that is, 70 to 80 feet—a full 2 to 2½ inches is used. The battens must have their edges chamfered. The solid ceiling is normally fastened with flathead galvanized screws so that it may be removed easily. If explosives are to be carried, fastenings are counterbored and plugged and all other metalwork is covered to avoid the possibility of sparks.

Ladders and Passageways

The minimum width for companionway ladders is 16 inches over the rails. There is a tendency to make them as wide as 30 inches, but wide ladders become dangerous in a seaway. The minimum width of an enclosed stairwell is 23 inches. The minimum width of an opening, such as one between two counters, is 19 inches if it is not above waist height. The minimum width for a longitudinal passageway is 28 inches, and thus if port and starboard cabins are desired, each having only a single berth width, the minimum possible beam is 11 feet; 16 feet of molded beam would be considered a more normal minimum to avoid a boxlike vessel. There is a great loss of space in passageways, just as there is with every door that pierces a bulkhead, since space is lost on both sides of the bulkhead. The notion that, for reasons of safety, one should be able to pass from one end of the vessel to the other without going on deck has no

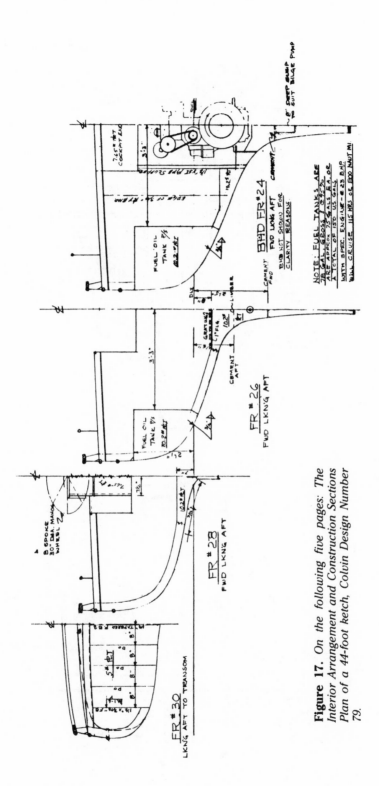

Figure 17. *On the following five pages: The Interior Arrangement and Construction Sections Plan of a 44-foot ketch, Colvin Design Number 79.*

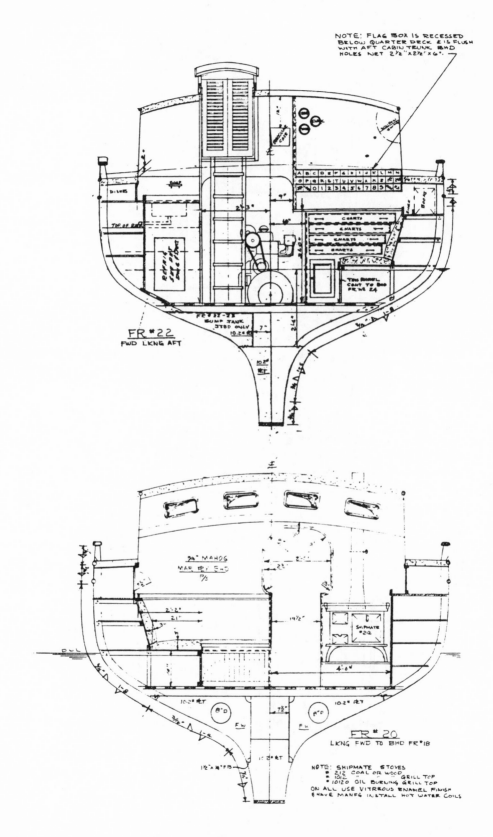

NOTE: FLAG BOX IS RECESSED
BELOW QUARTER DECK & IS FLUSH
WITH AFT CABIN TRUNK BHD
HOLES NET 2½" x 2½" x 6".

FR #22
FWD LKNG AFT

FR #20
LKNG FWD TO BHD FR #18

NOTE: SHIPMATE STOVES
  # 212 COAL OR WOOD
  # 10½C   GRILL TOP
  # 10120 OIL BURNING GRILL TOP
ON ALL USE VITREOUS ENAMEL FINISH
& HAVE MANFG INSTALL HOT WATER COILS

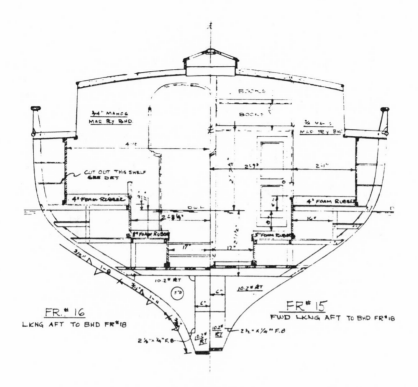

FR.# 16
LKNG AFT TO BHD FR#18

FR#15
FWD LKNG AFT TO BHD FR#18

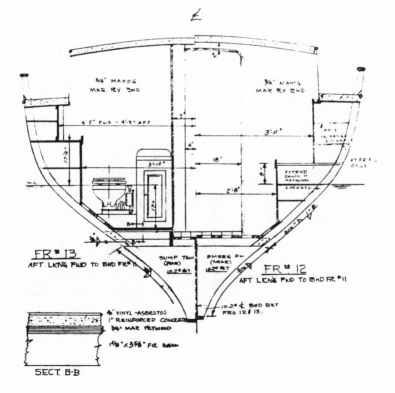

FR.# 13
AFT LKNG FWD TO BHD FR#11

FR.# 12
AFT LKNG FWD TO BHD FR#11

SECT B-B

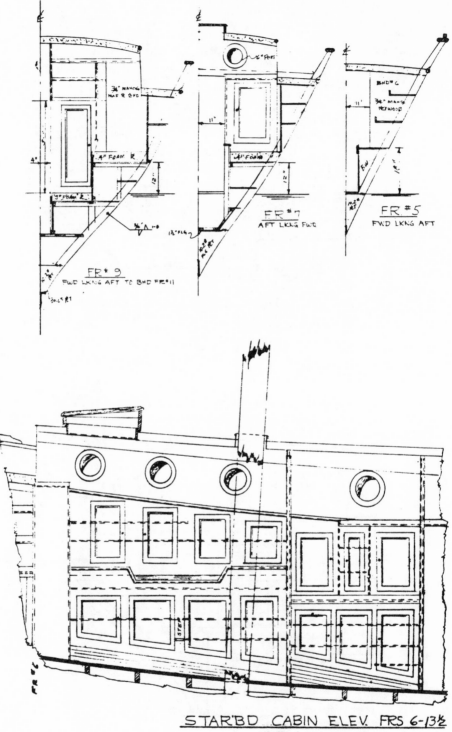

FR # 9
FWD LKNG AFT TO BHD FR # 11

FR # 7
AFT LKNG FWD

FR # 5
FWD LKNG AFT

STAR'BD CABIN ELEV. FRS 6-13½
INBD LKNG OUT B'D

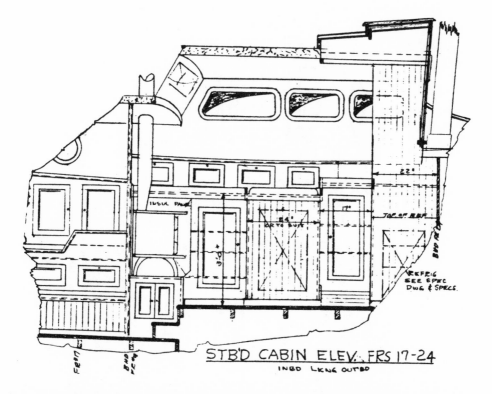

STB'D CABIN ELEV., FRS 17-24

INBD LKNG OUTBD

basis in fact, but is a gimmick to sell yachts. These remarks are addressed as much to those who are contemplating modifications to their vessels' interiors as to the builder of a new vessel. A study of Figure 17 will give one an idea of spatial arrangements in a 44-foot ketch, and the dimensions that proved to be convenient for the owner in a seaway.

EXTERIOR JOINERWORK

As has been mentioned, the exterior joinerwork on a steel vessel can be quite extensive, and with a few exceptions, the long-term maintenance problems presented by this joinerwork have not yet been solved. It is true that there are numerous short-term solutions, but in considering these, the builder and owner should decide whether longevity or adherence to a fad is the requirement. This is particularly true where wooden decks and railcaps are concerned, and deck joinery is further complicated when a builder tries to overlay a steel deck with a layer of wood that is not thick enough to be caulked, counterbored, and bunged. This thin wooden decking requires numerous blind fastenings from below, through the steel, but this practice should be avoided for several reasons: the fastenings into the wood are heat transfer points that will permit the steel to rust and the wood to rot as moisture seeps through; disintegration of the fastener, along with the fact that the wood does not have a second free surface from which to expel moisture and

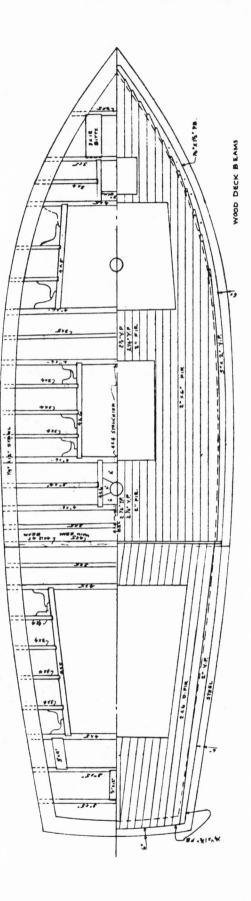

Figure 18. *Above and below: The deck plans of two sister ships, small coasting schooners. In the one below, 3½-inch by 2½-inch by ¼-inch angles are used for deck beams. The other uses wooden beams (3 x 4s, 3 x 5s, 4 x 5s, and 4 x 6s) reinforced by lodging knees and pillars.*

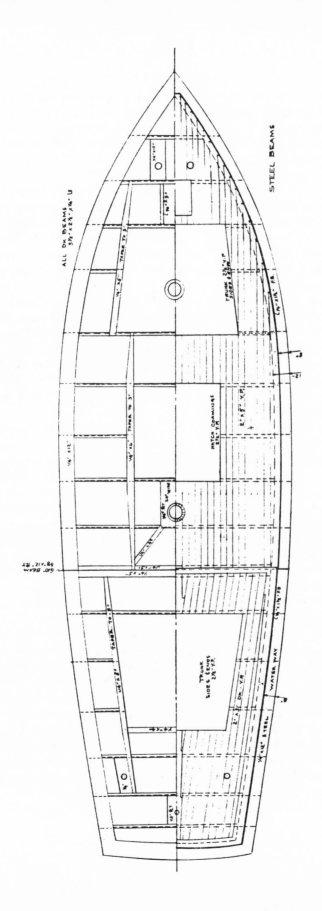

therefore prevents the steel from drying, will together cause poultice corrosion; and the bedding compounds used to separate the wood from the steel can be perfectly laid in theory, but seldom are in reality. Some designers and builders have substituted a plywood underlayment in lieu of a steel deck, and then laid the finished deck onto this by blind fastening from below. Again, in the long term, the whole deck will rot and the steel beams will rust.

About the only chance one has of laying a wooden deck on steel deck beams is to use the proper thickness of wood to begin with, and to bolt this wood through each deck beam, luting each faying surface. The decks are then caulked in a proper fashion. Eventually, of course, there will be a deterioration of the steel beam, but this will ordinarily occur perhaps 20 or so years after building. The preferred method is to use wooden deck beams that are in turn bolted to the frames, thus confining the problems to one small area.

No matter how a wooden deck is laid on a metal vessel, there must be a waterway plate between the edge of the wood and the deck edge that remains as exposed steel. In the way of cabin trunks, hatches, and other deck structures, the joints of wood to steel are quite complicated if they are done correctly. Wooden decks less than 1½ inches thick are a waste of time and money. Figure 19 shows the proper seam in a 2-inch wooden deck. Glue and tar (pitch) are the usual seam fillers. Single and two-part polysulfides are used in yachts. No filler is permanent, and all wooden decks will leak eventually, usually over one's berth.

Figure 18 shows the deck plan of a small coasting schooner, in which one may see steel deck beams and the waterways and other reinforcements necessary to lay a wooden deck. Also shown is a sistership in which wooden deck beams were used; note the extensive use of lodging knees required when steel beams cannot be used. This vessel required stanchions (pillars) not only under the hatch beams but also under the mainmast partner beam. The scantling difference between the steel and wood versions indicates the loss of depth due to wooden beams.

Figure 19 shows many of the details that are used when decks are of wood. Also shown is a wooden cap rail. No satisfactory long-term solution exists to avoid the rotting of wood and corrosion of steel associated with a wooden cap rail; therefore, the use of steel tubing that has been flattened is to be preferred.

There is one wood-to-steel joint above the weather deck that seems to be acceptable, that being the joint between a boundary bar and the top edge of the cabin trunk (but only when there is some overhang, so that water cannot settle in the joint). The success of this joint depends on drilling all holes prior to sandblasting, painting the surface properly with both the primer and the barrier coats, avoiding destruction of the barrier coats that line the holes when drilling the wood for bolts, and using a polysulfide for the luting. When bolting the marine ply cabintops in place, the builder pulls the heads of the carriage bolts into the wood, and after all the bolts are tightened, he uses a 3-pound maul to hit a rounded bronze dolly bar, driving the heads below the surface of the wood. The nuts are then tightened and the heads covered over with an epoxy paste. After the paste is cured, the whole is covered with epoxy resin and lightly sanded, after which it is covered with fiberglass cloth laid in epoxy resins. Wetting out fir plywood only from the top of the cloth will result in poor adhesion in some instances; hence, the prior coat of resin is necessary.

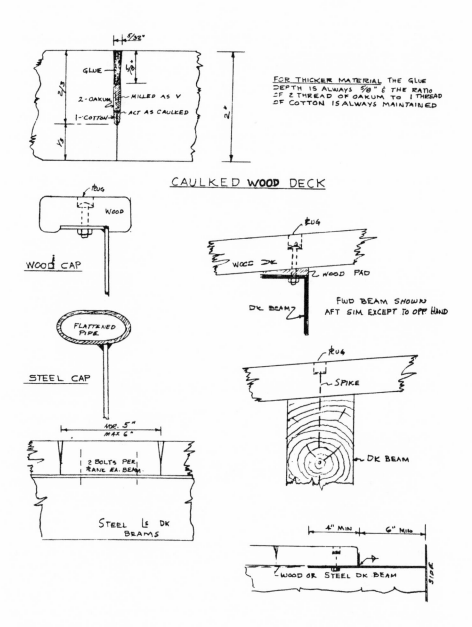

CAULKED **WOOD** DECK

FOR THICKER MATERIAL THE GLUE
DEPTH IS ALWAYS 5/8" & THE RATIO
OF 2 THREAD OF OAKUM TO 1 THREAD
OF COTTON IS ALWAYS MAINTAINED

Figure 19. *Some details of wooden deck construction, including a proper seam in a 2-inch wooden deck, a waterway plate inside the deck edge, and a wooden versus a metal cap rail.*

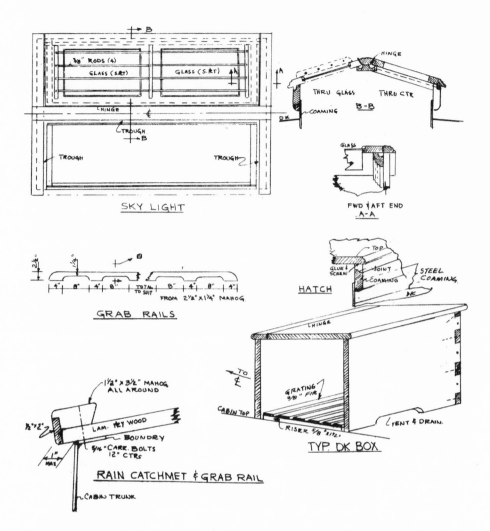

Figure 20. *Details of a typical skylight, a hatch coaming, a deck box, a cabintop edge molding, and grabrails.*

Figure 20 shows a cabintop edge molding that is both a grabrail and a rain catchment; it has been standard on my commercial designs and in my yard for the past 30 years. The catchment can be built of almost any wood except teak, which stains the water. Also shown is a section through a hatch, the construction of which not only assures proper maintenance of the steel deck-to-coaming joint, but also incorporates a wood-to-wood joint that will not leak when the hatch is dogged down. The corners of wooden hatch coamings and other low deck structures are a source of trouble if the coaming tops are too close to the deck, as there is then insufficient wood to secure them in all directions. End grain must never be exposed in such a way that water will lodge against it, for it will soon rot. The corner shown is a good one, requiring less material and time than a cornerpost would.

At best, a traditional skylight requires considerable time to construct and finish. All skylights leak, which, of course, can be stopped with a proper canvas cover. If this cover is of natural or off-white color, quite a bit of light will penetrate below when the hatch is closed. The cover also allows the wings to elevate to catch some wind, and it does save the finish of the skylight hatch. A typical skylight is detailed in Figure 20.

Grabrails are best constructed of pipe, a task within the realm of the builder. After fabrication, they may be sent off and galvanized or may receive the same finish as the vessel. If wooden ones are required, the ones detailed in Figure 20 have proven both strong and convenient.

Good-quality cast bronze, polished bronze, and polished brass hardware have become increasingly difficult to find. The reasons given are that too much labor is involved in the casting, that the wages one would have to pay just to have someone polish it daily would be too high, and that brass and bronze have gone out of style. All this may be true, but unpolished metal is more sightly than the alternative, which is flaking chrome plating. The amount of hardware needed is more or less up to the builder and owner. If good joinerwork is available, wooden latches, hinges, pulls, and knobs can be made. There is also the possibility of using leather hinges and pulls. It seems that the least amount of hardware is the best, whether it be on the interior or exterior. The prerequisite for all hardware is that it be strong enough for the job it is supposed to do, but not oversized. In some vessels, the hardware appears as ridiculous as a twentypenny nail used to secure ¼-inch plywood.

CONCLUDING REMARKS

Detailed descriptions of how to construct all the joinerwork on a vessel are beyond the scope of this book. The builder would be well advised to consult one or more of the numerous books on furniture making and cabinetmaking, and even the "how to" books developed for yachtsmen. Most of the problems anticipated at the beginning of the job will automatically solve themselves as the work progresses. Joinerwork is not a race against time, for that approach will lead to mistakes that are costly not only in time but in material. If everything seems to be going wrong, then it is best to stop and do nothing *("Stand there, don't just do something!")* until the problem is thought out and a solution is apparent (which is usually in the middle of the night).

Be wary of laminating different colors of wood together. While it is necessary in a chess or backgammon board, it will look more like a bent barber pole than a work of art if used in beams and other members. Numerous gimmicks and gadgets are available through the various wholesale and retail outlets. Some of these are trash, and some can be fine additions or substitutions for certain joinery; however, if such an item can be added at a later date, the builder should by all means wait and determine that it is really necessary. Plywood makes fine cabintops, berth and settee bottoms, shelves, and drawer bottoms. It is fine also for cargo hold and engine room bulkheads. It is not cheap nor pleasant to work with, however, and it is certainly not attractive to look at over a period of years. As for any of the plastic coatings and laminates, one can soon tire of their sameness, but there is no question that they do have a place in the scheme of present-day living.

A cabin that is to sleep two, seat six, and contain the galley, a couple of hanging lockers, a stairway, and a water closet often occupies space smaller than a bathroom in a house ashore. Remembering this, a builder realizes the importance of careful planning and the subtle nuances that are so necessary in a livable interior.

3

▽

MASTS,
SPARS,
AND
RIGGING

Designing the sail plans for commercial vessels is not an exercise in trying to revive what our forefathers developed or to reinvent their rigs; rather, it is a process in which one accepts what was developed in the past and improves it whenever possible, especially when what was developed can be applied to vessels engaged in a similar occupation. In the commercial world, the key to survival is profit, not nostalgia. Working any vessel commercially requires low first cost, low maintenance, small crews, and easily handled rigs. Those who have never been shipmates with the numerous commercial rigs that exist today tend to ridicule and condemn them as being heavy, complicated, and inefficient. Yet how many sailors owning 50- to 60-foot vessels with so-called modern and efficient rigs would even contemplate a passage of a thousand miles or more, singlehanded or even with a crew of one, especially without an auxiliary to get them in and out of port and charge the batteries for the autopilot, Loran, SatNav, and so forth? The commercial sailor often makes such passages, using the old rigs and their modern adaptations because of their convenience for small crews, their ease of repair, and their versatility.

Many auxiliary cruising yachts also use some of the older, lower aspect ratios and gaff rigs for the same reasons as the commercial sailor. The ocean-cruising vessel is often sailed shorthanded and sometimes singlehanded. In cruising, the vessel's ability to work close to the wind and sail as fast as possible closehauled is important, but never more so than her overall excellence on the other points of sailing. When working off a lee shore in a gale, a vessel with adequate sail area depends as much on hull shape (her ability to carry sail) and a generous amount of lateral plane as she does on the efficiency of her sail plan. In tacking, she must not go in irons, nor make sternway in order to right her helm, nor sag off after tacking before filling her sails, thus losing what she has gained.

In this chapter, consistent stress will be placed on the less expensive ways of constructing masts, spars, and rigging, as well as the interrelationship of each part to other parts and to the rig as a whole. Many companies today will furnish "modern spars and rigging" to the builder as a package, and few builders wish to become involved in manufacturing what they can purchase as a more or less stock item. Most of the "modern" rigs, however, are highly stressed and are under great compression loads. With the exception of a few highly sophisticated yards, most builders must subcontract all such spar and rigging work, since they do not have the necessary extrusion and swaging equipment. When working with the older types of rigs, one can splice the rigging or use one of the patented splicing sleeves. One can also fabricate or manufacture the masts, spars, and fittings with the same equipment used to construct the hull.

I do not mean to suggest that because one is building a steel vessel, he should rig it like a cargo schooner or a fishing smack. However, a study of commercial rigs will indicate the simplicity required to handle large sail areas with a minimum of crew. Given some thought and ingenuity, the commercial rigs are inexpensive and require very little in the way of maintenance by the crews. Incorporating some of their details into a cruising yacht would certainly lessen its first cost. The details shown in this chapter, therefore, indicate some of what can be done with standard materials available from many sources or manufactured by the builder. Details so specialized that they may well be available only from one source are excluded. The builder or owner of a yacht intended solely for racing, which normally carries a large crew for her size and sail area, will learn little from this chapter. It would seem that fashion is of prime importance, and that, right or wrong, following the leader is the rule of the day in spite of the fact that every vessel has her own compass. For those who want them, photographic details of the latest trends in rigging can be found in the various yachting magazine articles.

Just a few decades ago the designer had only to design the sail plan for a schooner or any other rig, and indicate where the rigging went, and its size. The remaining details were so standard that it was assumed everyone knew all about them, and the builder, blacksmith, sparmaker, and rigger would automatically do what was needed. Today, the designer can no longer assume that this detailed knowledge is common. The following rigs are shown, then, because they encompass most of the standard rigging details needed on spars that can be constructed by the builder or the owner of a steel vessel.

SELECTED RIGS

The brigantine rig is seldom used these days because of the displacement needed to utilize the sail plan successfully. Vessels in the past that used this rig were, by today's standards, very burdensome—typically about 50 feet on deck, with 16 feet beam, 8 feet 6 inches depth of hold, and a displacement of 50 to 60 tons. These vessels spread close to 3,000 square feet of sail area, not including studding sails, water sail, ringtails, and numerous other light sails, yet were handled by a crew of four. The brigantine makes a very efficient rig for ocean voyaging, since the combinations of exposed sail area are easily adjusted instead of reefing, and the staying is excellent. Off the wind it performs well without the addition of extra sails. A brigantine is faster than a schooner of identical hull form, but is not as close-winded nor as easy to tack. An example of this rig is the brigantine *Down North*, built in Labrador using the same lines that were used for a small coasting schooner (Figure 21). Her dimensions are as follows: 45 feet on deck, 38 feet 6 inches DWL, 14 feet 2 inches beam, 6 feet depth of hold, 23 tons displacement, and 1,677 square feet of sail in the working rig.

Squaresails on brigantines are furled by going aloft in the normal manner, using ratlines in the rigging, and then out on the yard using footropes attached to the yard with stirrups. The sails are fastened to a jackstay fitted atop the yard. The fore yard is hung on a sling chain and attached to a swiveling truss, and the topsail rides up and down the fore topmast on parrells. On smaller vessels, the foresail clew lines and the topsail sheets lead to special double blocks on the lower side of the sling eyeband of the fore yard, and the topsail clew lines lead to the center of the topsail yard, all of which then lead to the deck, enabling one person to do most of the work on deck before going aloft to furl.

The use of square rigs requires that careful attention be paid to the location of shrouds and stays, so that they will not interfere with the yard when it is braced up sharp. This is especially true with courses, because these sails are fitted with a truss and therefore swing to leeward. There is a tendency nowadays to cut a deep roach in the foot of the topsail because, at least on the drawing board, this roach appears necessary for the topsail to clear the forestaysail and jibstays by a good margin. In practice, the fore topsail bellies out enough to give clearance, and chafe will only occur with the sail aback. Yards not only rotate around the mast but must also swivel so that they may be trimmed in the horizontal plane.

When a squaresail is used on a schooner that is not topsail rigged, it is usual to fit the yard to a vertical jackstay just forward of the foremast. This stay leads over a jumper strut located above the gaff jaws. The yard is secured with special fittings and can pivot as well as tilt. It is sent down when not in use, as the furled sail and yard create a lot of windage if the vessel's passage is dead to windward. On some vessels the squaresail is brailed in to the center of the yard, then further secured by wrapping the brailing lines about the sail in a spiral. When this is done, the use of a hoop or several hoops sliding out on each side of the yard in lieu of the jackline seems to be about the best arrangement to prevent jamming. This arrangement keeps the head cringles from drooping down when the sail is brought in. Sail track on yards has the disadvantage of jamming at rather awkward times, such as when the sail goes aback or during an attempt to set or furl sail going downwind. An alternative is to hoist the center of the head close up to the yard and then use outhauls to set the sail along the yard, leaving the head flying, port and starboard. This method has the advantages that it eliminates the need to go aloft; the yard may weigh less; and, if a crossleech is sewn to the sail, the lee clew may be hauled up and the midship tack hauled aft so that the foresail will not be backwinded. This makes an excellent downwind rig, easier to use than the typical tradewind spinnakers and less violent in rolling. In spite of the work associated with the squaresail on schooners, it does a magnificent job from close-reaching to running, and it is docile and easy to sheet. The normal procedure when the yard has been sent down at sea is to rehoist it in

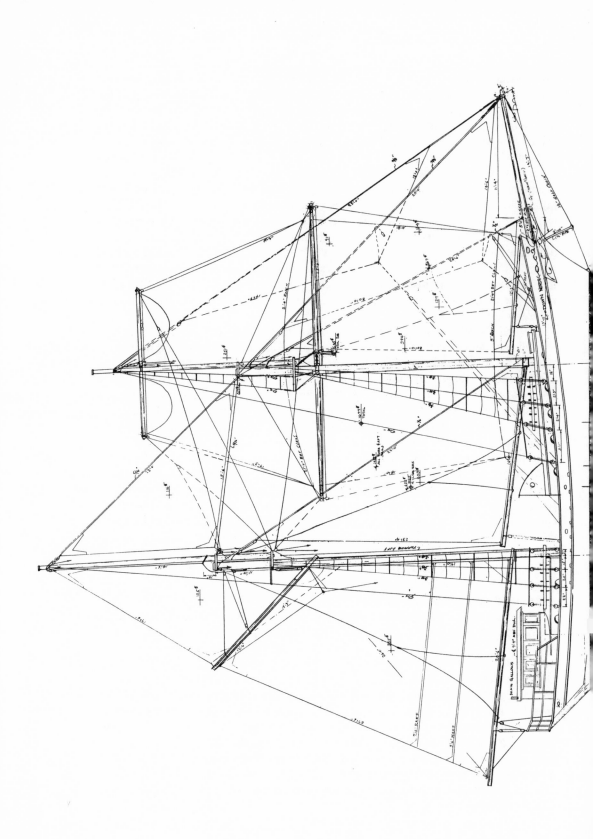

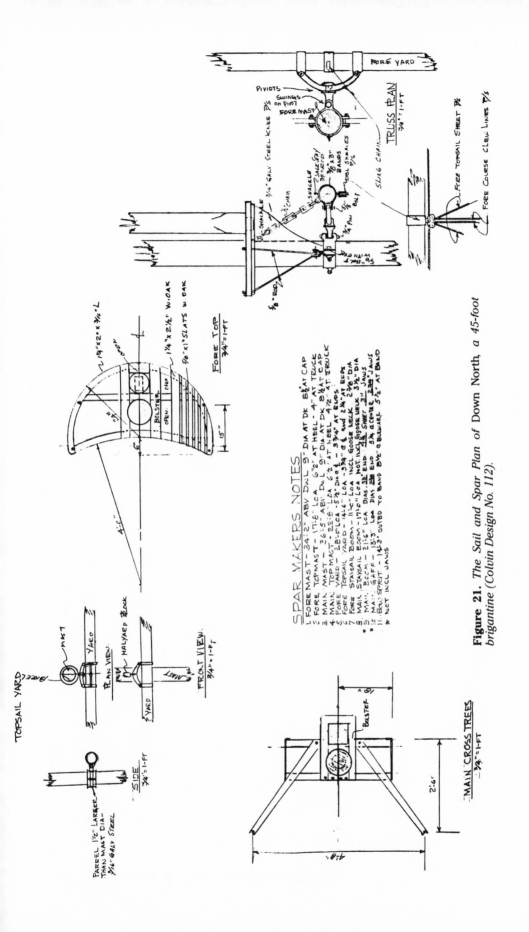

Figure 21. *The Sail and Spar Plan of Down North, a 45-foot brigantine (Colvin Design No. 112).*

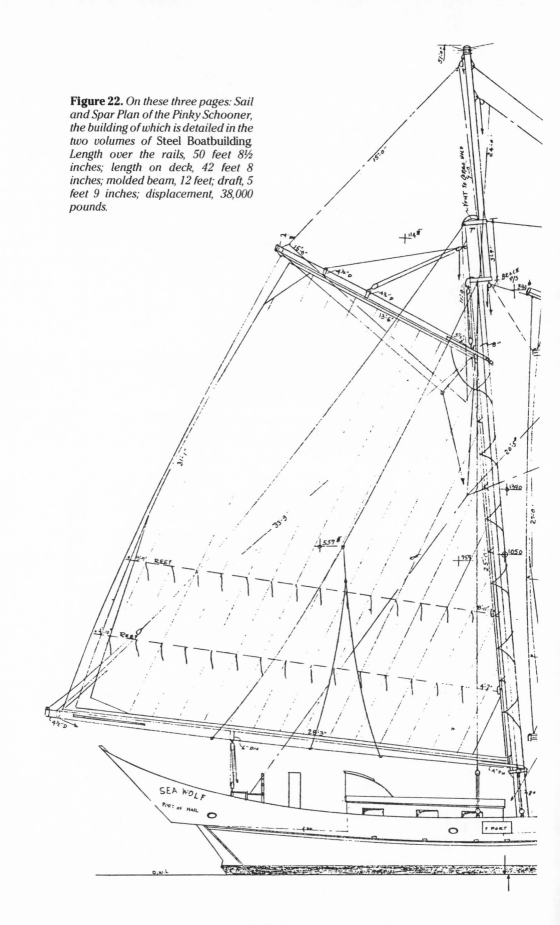

Figure 22. *On these three pages: Sail and Spar Plan of the Pinky Schooner, the building of which is detailed in the two volumes of* Steel Boatbuilding. *Length over the rails, 50 feet 8½ inches; length on deck, 42 feet 8 inches; molded beam, 12 feet; draft, 5 feet 9 inches; displacement, 38,000 pounds.*

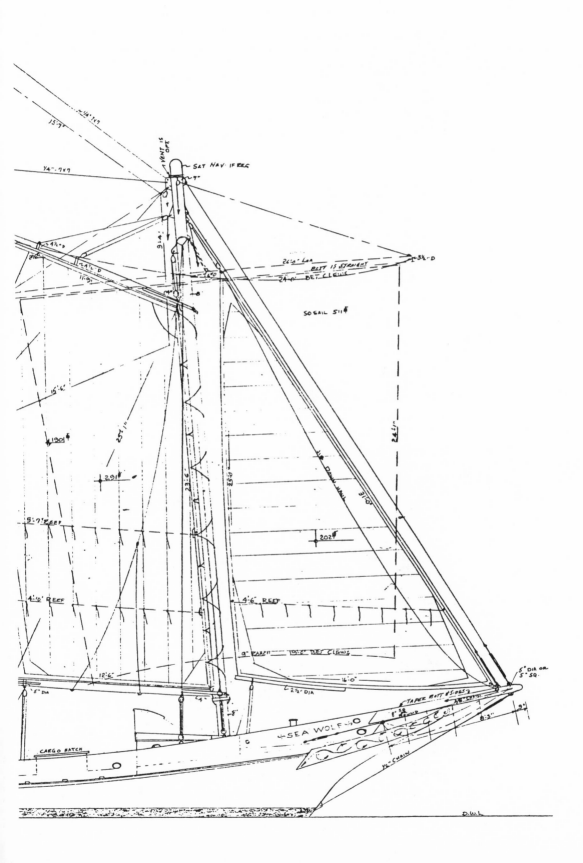

SAIL MAKERS NOTES:

1. SAILS TO BE TRIPLE STITCHED 13⁹ VIVATEX OR 7.75-⁴³ DACRON
2. SAILS TO HAVE HAND SEWEN BOLT ROPES ALL AROUND ~ JIB WILL HAVE A ROPE LUFF
3. DIMENSIONS ARE PIN TO PIN FULLY STRETCHED ~ MAST HOOPS ARE 11" DIA IF USED OR LACING AS SHOWN - OWNERS OPTION
4. N8 BATTENS IN ANY SAIL
5. CLOTHS ARE FALSE SEAMED IF VIVATEX OR 18" WIDE DACRON CLOTH TO BE USED
6. TOP SAILS TO BE 5-⁴³ DACRON
7. SQ. SAIL TO BE 6-⁴³ DACRON
8. REEFS ARE TO BE IN BANDS & NETTLES FITTED
9. GROMMETS AT EACH SEAM ON FOOT & HEAD.

SPAR MAKERS NOTES:

1. MASTS MAY BE OF WOOD, ALU. ALLOY OR STEEL TUBING - BOOMS & GAFFS ARE WOOD OR ALU. ALLOY.
2. MAIN MAST - IS 40'.5" ABV D.WL & 43'-9" LOA ~ 8" SCH 40 ALU. ALLOY PIPE (6061-T6), 9" DIA DOUG FIR, OR 8" STEEL TUBE 10 GA. WALL THICKNESS
3. MAIN TOP MAST 14'-0" LOA ~ 4" SCH 40 ALU. ALLOY PIPE OR 4¾" DIA. DOUG. FIR.
4. MAIN BOOM - 29'-7" LOA TO MAST - DEDUCT FOR CLAPPER ~ DOUG. FIR
5. MAIN GAFF - 14'-3" LOA TO MAST - DEDUCT FOR CLAPPER ~ DOUG. FIR
6. FORE MAST - 39'.5" ABV D.WL - LOA IS LESS AS. PER MAST STEP
7. FORE BOOM - 13'-2" LOA TO MAST - DEDUCT FOR CLAPPER ~ D.FIR
8. FORE GAFF - 12'-4" LOA TO MAST - DEDUCT FOR CLAPPER ~ D. FIR
9. JIB CLUB - 7'-0" LOA ~ DOUG. FIR
10. BOWSPRIT - 8" X 8" X .188 STEEL TUBE OR 8" DIA STEEL TUBE 10. GA MIN 7 GA. MAX
11. YARD - 6" STEEL TUBE - 11 GA - TAPERED
12. WISKER - 5'-0" LOA. 1½" SCH. 80. PIPE ~ END CLIPS 1" X 1¼" F. BAR
13. THUMBS - EYES - ET. & ALU. FT (ALU. SPARS) STEEL FT (STEEL SPARS): GALV. STEEL EYE BANDS ON WOOD SPARS
14. JAWS DETAILED ON DWG 35-B

EQUIPMENT:

1. 2 - 55# CMC - FISHERMAN TYPE - BOWERS. ANCHOR
2. 1 - 35# CMC - FISHER MAN TYPE - KEDGE ANCHOR
3. 50 FATHOMS 5/16" GRADE 40 ARCO CHAIN - STBD - WORKING BOWER
   45 FATHOMS 5/16" GRADE 40 ARCO CHAIN - PORT ~ CORAL - SPARE BOWER
   5 FATHOMS 5/16" GRADE 40 ARCO CHAIN - PORT + ¾" NYLON - BAUES ±
4. LIFE JACKETS FOR EA. BERTH + 2 EXTRA
5. MARINASPEC TRI COLOR LIGHT ON MAIN TOP MAST - M - 7001.
6. ANCHOR LIGHT (OIL)
7. 6" FOG BELL ~ MECH. FOG HORN
8. BILGE PUMP (MANUEL) 2" SUCTION & DISCHARGE (EDSON)
9. SESTRAL 5" SPHERICAL COMPASS FOR STEEL HULLS
10. 12' BOAT HOOK ~ GALV HOOK
11. 4½" SOUNDING LEAD & MARKED LINE (10. FATH)
12. 300,000 C.P. PORTABLE LIGHT ~ 2 OUTLETS
13. 6 MOORING LINES - ¾" NYLON - 8 FATH EA - WITH A 20" EYE SPLICED IN ONE END.
    3 WARPS ~ 50 FATH EACH ~ ¾" NYLON - SPLICE ON EACH END
14. AWNINGS & SAIL COVERS TO SUIT OWNER REQ
15. NAME & PORT OF HAIL ~ EA SIDE OF STERN & NAME ON EA. SIDE OF BOW.
16. 9' BOTT. LENGTH DORY - 3'11" BEAM 18" DEPTH ~ WOOD
17. 2 - DRY CHEM FIRE EXT (SAILING VESSELS
18. ANY & ALL OTHER EQUIPMENT REQ BY USCG OR FLAG OF REGISTER & REQ FOR SEA KEEPING ABILITY & SAFETY
19. LUNENBURG RATCHET GYPSY WINDLASS SIZE #1 FOR CHAIN ~ 2 WILDCATS
20. " " " " " SIZE #1 FOR ROPE & 2 WARPING HEADS

RIGGERS NOTES

1. STANDING RIGGING TO BE 7 X 7 GALV. IMPVD PLOW STEEL ~ SHROUDS & JIB STAY 3/8" DIA. HEAD STAY 5/16" DIA ~ TRIATIC & SPRING STAY ¼" DIA
2. ALL TERMINALS TO BE NICOPRESS SPLICING SLEEVES OVER THE PROPER SIZE SOLID THIMBLE WHEN REQ.
3. EYES OVER MASTS TO BE WORMED, PARCELED & SERVED
4. SHROUDS ARE PASSED OVER AS PAIRS FROM 1 CONT. WIRE ON EA. SIDE & ARE MARRIED WITH A NICO PRESS SLEAVE
5. DEADEYES FROM A. DAUPHINE & SONS - LUNENBURG N.S.
6. TURNBUCKLES ARE JAW & JAW - GALV. BARREL TYPE WITH LOCKING NUTS
7. BOB STAY IS ½" GALV. CHAIN - BOWSPRIT SHROUDS ARE 3/8" GALV CHAIN
8. LAZY JACKS ARE 5/16" (3) STRAND DACRON, & MAY BE SET UP BOTH F/S.
9. TOPPING LIFTS ARE 3/8" (3) STRAND DACRON
10. HALYARDS FOR LOWERS & THEIR SHEETS IS ½" (3) STRAND DACRON
11. TOPSAIL SHEETS, HALYARDS, AND SQUARE SAIL LIFTS, BRACES, & SHEETS 3/8" (3) STRAND DACRON ~ OUT HAULS, IN HAULS, & CLEW LINES (7) (3) STRAND DACRON
12. JIB DOWNHAUL ~ 5/16" DACRON.
13. PIN RAILS P/S FORE & MAIN RIGGING
14. FORE & MAIN SHROUDS ARE RATTLED DOWN.
15. JIB MAY BE RIGGED BUGEYE STYLE ~ OWNERS OPTION.

port. At a dock, it is braced around; at anchor, it is braced square without the sail bent, since the yard is then out of the way, leaving the rails clear.

Square-rigged vessels close-hauled cannot lay closer than 6 points to the wind (67½ degrees). In fact, many of the full-rigged ships and barks cannot lay closer than 7 points. Around the buoys, this would be a catastrophe; in the open ocean, however, it is of small consequence. The schooners lay 4 points to the wind and, if fine-lined, will also make it good. The squaresails are seldom used when beating to windward for just a few hundred miles. As the speed of a vessel increases, her ability to point also increases; however, to strap a vessel in just to lay closer to the wind is futile if, at the same time, her leeway increases. It is the course made good that is important, and not how close she can lay.

The Pinky Schooner illustrating this book is a two-masted, gaff-rigged vessel, with her yard permanently set aloft due to the waters in which she was designed to sail. If she were used exclusively in northern waters rather than in the trades, the yard would either have been dispensed with or set on a jackstay. Not having an engine, she is blessed with a generous sail area: 1,050 square feet in the three lowers, 1,390 square feet in the fore-and-aft rig, and a total area of 1,901 square feet. Her squaresail is set flying, eliminating the need to go aloft to furl. Because she carries her workboat on deck along with deck cargo, her jib sheet leads bugeye fashion to the underside of the truss and thence to the railcap, freeing the decks of all running rigging. In spite of all the ill heaped upon not only the gaff rig but upon schooners as well by writers having limited or no experience at sea with the rig, it is a fine seagoing rig that can spread an enormous amount of sail on short spars. One can easily add sail, thereby increasing the total area, rather than having to follow the modern practice of substituting one sail for another. After all, it is sail area that propels the vessel, and in light weather generous amounts are necessary to keep her moving. A bowsprit, often unjustly called a "widowmaker," is an inexpensive addition to the hull and, with a safety net below, provides a fine place to work. It is better than a platform with a pulpit, and it aids in keeping the center of effort low. It is also such a lever in relation to the hull that it does a magnificent job when the sails are trimmed for self-steering. In commercial vessels, the headsails are always fitted with downhauls, which relieve the crew from having to go out on the bowsprit to muzzle the jib.

While I was in Martinique, I prepared the design shown in Figure 23 for two fishermen, one French and one Belgian. The two vessels were to be used as trollers, handliners, and longliners; hence the round stern. The rig is of Bahamian origin, with the solid headboards made of plywood. Had short gaffs been used, then the origin would have been Dutch. The loose foot is common to many European types, but the exaggerated round to the foot is Bahamian. The fore and main are hoisted with single halyards, and the jibsheet is fitted bugeye style. This has proven a handy rig for fishing, since the loose-footed sails can be detached from the booms and sheeted just from the clews, and the booms are then elevated and used as outriggers, thereby eliminating having to contend with extra poles while fishing and sailing. For tradewind sailing, one could possibly object to the high aspect ratio of the rig when running and to the use of sail track on the mast instead of lacing, which prevents lowering sails on the run. The masts were specified as aluminum alloy but were built in steel. These vessels measure 52 feet 6 inches on deck, 41 feet 7 inches DWL, 13 feet

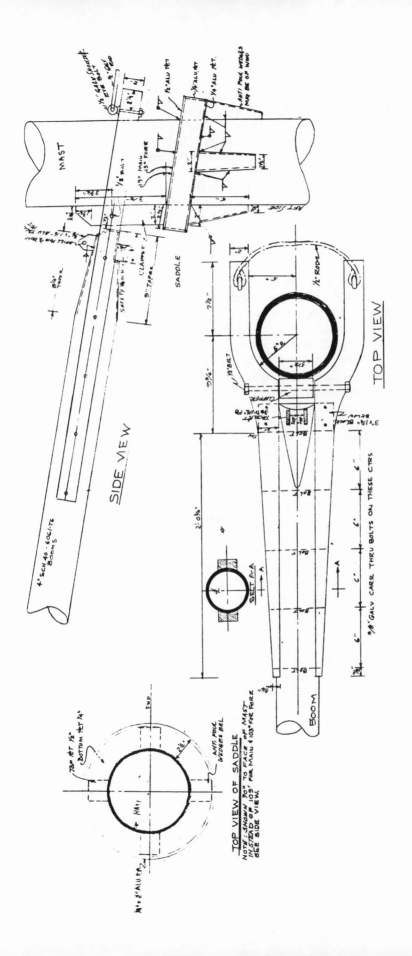

Figure 23. On this and the three following pages: Sail and Spar Plan of a 53-foot cargo and fishing schooner, Colvin Design No. 184.

SAIL MAKERS NOTES

1. SAILS TO BE TRIPLE STITCHED 13% VIVATEX OR 7.75% DACRON.
2. SAILS TO HAVE HAND SEWN BOLT ROPES ALL AROUND - FOOT OF FORE & MAIN SAILS TO HAVE 5/8" DIA (MIN) ROPES & NORMAL SIZES ELSEWHERE
3. DIMENSIONS ARE PIN TO PIN. FULLY STRETCHED - LESS ALLOWANCE FOR CUT BACK OF 3" ON MAIN & FORE TACKS TO FIT TACK PIN - CURVING TO FIRST SLIDE 24" UP
4. HEAD BOARDS ARE 3/4" MAHOG MARINE PLY & ROPING TO BE CARRIED OVER & UNDER (ALL AROUND) SAILS & LACED TO HEAD BOARD WITH 20 THREADS OF SAIL TWINE.
5. REEFS ARE TO BE IN BAGS & KETTLES FITTED
6. LAZY JACK THIMBLES ARE 1" ID BRASS & SEWN PARALLEL TO BOLT ROPE
7. CLOTHS ARE 18" OR FALSE SEAMED 1" VIVATEX
8. FOOT ROUND TO BE SCALED FROM SAIL PLAN
9. NO BATTENS IN FORE OR MAIN SAILS

SPAR MAKERS NOTES

1. ALL SPARS TO BE 6061-T6 ALUMINUM ALLOY TO DIAS GIVEN & INDICATED.
2. MAIN MAST - 59'0" LOA - LOWER 24-4"-8 SCH 40 PIPE - UPPER 34'-6 - GRAYBAR SECT 6.1-.088 1/4" WALL TAPERS 8" TO 4.5" (478")
3. MAIN BOOM - 23'-6" LOA FROM AFT SIDE OF MAST (DEDUCT CLAPPER) 4" SCH 40 PIPE
4. FORE MAST - 57'2 LOA - LOWER 22-6"-8 SCH 40 PIPE - UPPER 34'-6 - GRAYBAR SECT 6.1-.088 1/4" WALL TAPERS 8" TO 4.5" (458")
5. FORE BOOM - 17'9" LOA FROM AFT SIDE OF MAST (DEDUCT CLAPPER) 4" SCH 40 PIPE
6. BOW SPRIT - 8-11.5' LOA FROM RAIL (SEE SAIL PLAN) MADE FROM 10 GA PLT BOX SECT
7. ALL SPARS TO BE ETCHED & PAINTED - BOW SPRIT IS SAND BLASTED & PAINTED
8. THUMBS (FOR SHROUDS) BLOCK PLATES - MAST HEAD & BOWSPRIT FITTINGS ARE TO BE OF SIMILAR MATERIAL AS THEIR SPAR
9. MAIN & FORE MAST FITTED WITH 7/8" SAIL TRACK - MOUNTED ON 5/8" THICK PURING STRIP TO PERMIT SHROUDS TO PASS AROUND MAST
10. MAIN SADDLE IS 46-8" FROM MAST HEAD ~ FORE SADDLE IS 46'-4" FROM MAST HEAD

EQUIPMENT:

1. 45# COR PLOW
2. 60# COR PLOW
3. 15 FATHOMS 3/8" BBB GALV CHAIN + 45 FATH 3/4" NYLON FOR #1 } ANCHOR 15 FATHOMS 3/8" BBB GALV CHAIN + 60 FATH 3/4" NYLON FOR #2 } RODES
4. LIFE JACKETS FOR EA. BERTH (USCG OR BOT TYPE)
5. MARINASPEC TRI-COLOR ON MAIN MAST TOP # M7001
6. ANCHOR LIGHT (OIL)
7. FOG HORN & FOG BELL
8. BILGE PUMP (MANUAL) 2" SUCTION & DISCHARGE (EDSON)
9. SESTRAL 5" SPHERICAL COMPASS FOR STEEL HULLS
10. 10' BOAT HOOK
11. 4.5# SOUNDING LEAD & MARKED LINE (10 FATH)
12. 300,000 CP PORTABLE LIGHT
13. 6 MOORING LINES - 3/4" NYLON - 8 FATHOMS EA WITH 20" EYE SPLICE IN ONE END & SEIZING ON OTHER END
14. AWNINGS & DODGERS & SAIL COVERS TO SUIT OWNER REQ
15. NAME & PORT OF HAIL ON STERN - NAME ON EA. SIDE OF BOW
16. 10' DINGHY ON DECK - T.E.C. STD OR EQUAL
17. 2 DRY CHEMICAL FIRE EXT (SAILING) OR 3-MOTOR
18. ANY & ALL OTHER EQUIPMENT REQ BY U.S.C.G OR FLAG OF REGISTER & REQ FOR SEA KEEPING ABILITY & SAFETY
19. LUNENBURG RACHET GYPSY WINDLASS SIZE #2 ~ WITH 2 WILDCATS ONLY
20. LUNENBURG "PURITAN TYPE" STEERER SIZE NR 1 (32" OR 36" WOOD WHEEL)

RIGGERS NOTES:

1. STANDING RIGGING TO BE 7X7 GALV. IMPVD PLOW STEEL OR S. STEEL SHROUDS & JIB STAY 3/8" & TRIATIC & HEAD STAY 1/4"
2. ALL TERMINALS TO BE "NICOPRESS" SPLICING SLEEVES. OVER THE PROPER SIZE THIMBLE WHEN REQ
3. EYES OVER MASTS TO BE WORMED PARCELED & SERVED
4. UPPER SHROUDS ARE SINGLE ~ LOWER SHROUDS ARE PASSED OVER IN PAIRS PPL FROM ONE CONT. WIRE & ARE MARRIED WITH A "NICOPRESS" SLEEVE
5. TURNBUCKLES ARE 5/8" JAW & JAW GALV. BARREL TYPE WITH LOCK NUTS
6. BOB STAY IS 2" GALV. CHAIN
7. WISKER SHROUDS ARE 3/8" GALV. CHAIN
8. LAZY JACKS ARE 5/16" (3) STRAND DACRON & ARE CONTINUOUS LINES - FORE & MAIN CAN BE SET UP P/S. - JIB AS PER DWG.
9. TOPPING LIFTS ARE 3/8" (3) STRAND DACRON.
10. HALYARDS & SHEETS TO BE 1/2" (8) STRAND DACRON.
11. JIB DOWN HAUL TO BE 5/16" DACRON 3 STRAND
12. BLOCKS TO BE GIBBS OR EQUAL
13. HALYARDS, LIFTS & LAZY JACKS BELAY TO PIN RAILS P/S WHICH IS ATTACHED TO THE SHROUDS - HALYARDS MUST PASS FIRST TO A SNATCH CLEAT OR SNATCH BLOCK FITTED TO THE RAIL CAP FOR SWAYING UP & FOR A DOWNWARD LOAD ON THE PIN RAIL
14. JIB SHEET IS RUN AFT VIA FAIR LEADERS ATTACHED TO RAIL CAP ON PORT SIDE ONLY.

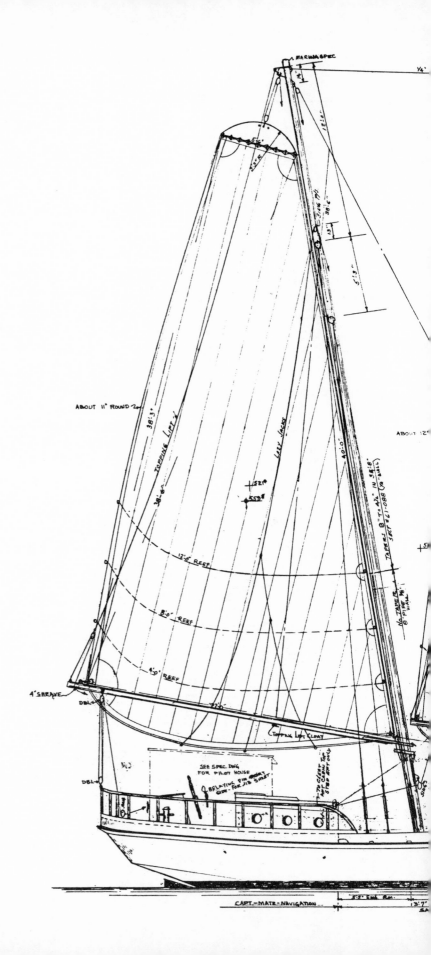

CAPT—MATE—NAVIGATION

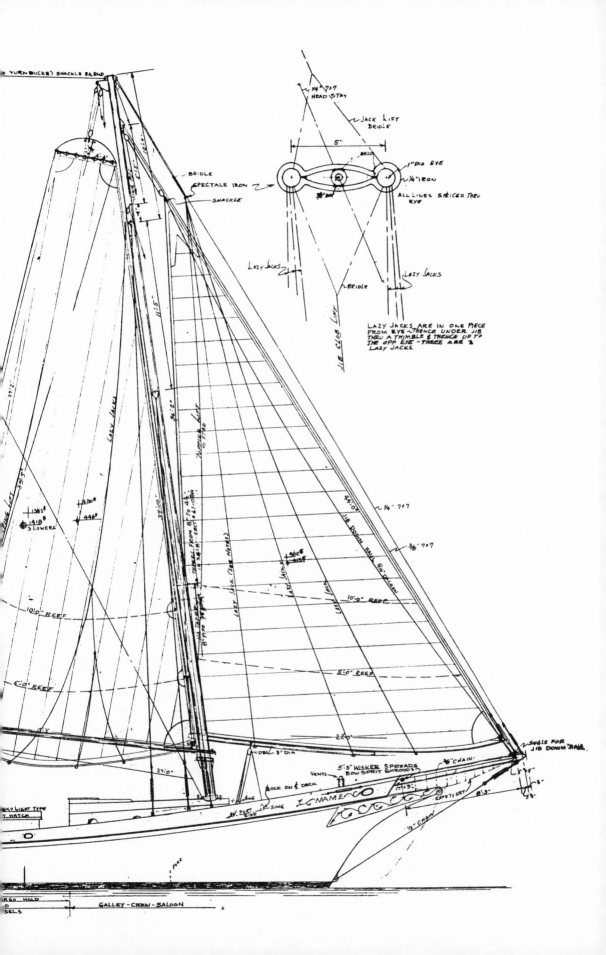

7 inches beam, 6 feet depth of hold, 1,478 square feet working sail area, and 16 tons displacement.

The trading schooner *Phantom* (Figure 24) utilized the Bahamian headboards to maximize her sail area without excessive mast height, which was limited to 48 feet, but all else is pure bugeye rig. The strongly raked masts with sails fastened to hoops or lacings allow her to lower sails downwind ("blow them down"). The rig's only disadvantage is that in light weather, on a run, the booms need preventers to keep them from coming inboard. In heavily raked rigs, the after shrouds are as much backstays as they are shrouds. Such rigs also make it easier to keep the rig set up, since only medium tension will suffice on the turnbuckles or deadeyes. *Phantom* carries her motorized workboat over the stern in davits, almost by necessity, because her boom overhangs the stern by 10 feet 6 inches, and the sail would otherwise be difficult to reef without the use of footropes. She carries two more workboats on deck, used for loading and off-loading cargo in some of the remote ports lacking dock facilities. Technically, one might call *Phantom*'s rig a ketch, since the mainmast is shorter than the foremast, but the mast placement is that of a schooner and her main boom is 3 feet longer than the fore boom. The advantage of a shorter mainmast is that there is little tendency for the main to blanket the foresail when running. The top of the foremast is directly above the cargo hatch, allowing use of the fore halyard for direct lifting of cargo. This vessel was designed for light packaged freight and so had no need for the heavy lifts that some small general freight schooners find necessary. If a heavy lift were needed, the fore boom could be topped up and rigged for it. All sails have lazyjacks. Those on the jib are attached Chesapeake Bay fashion, with the lazyjacks made up on a spectacle iron mounted on the headstay. Having the jib fitted with lazyjacks means that, when lowered, it will not land on the deck or in the water. Lazyjacks never eliminate the necessity for using a downhaul on the jib, although the use of just a club does eliminate the need for a jackline fitted to the luff of the jib. Lazyjacks are not fastened where they pass below a boom or the foot of the sail. They are made with bights of line and thus are self-adjusting and free to belly with the sail on the lee side, preventing chafe. Under the booms, eye straps are used; under the foot of the jib, grommets are worked in below the boltrope with their eyes open athwartship. The main staysail is of an unusual cut and in practice is more powerful than a triangular staysail that tacks to the deck. The open rail over low bulwarks is for safety in a seaway, permitting the quick freeing of water coming on deck. Solid bulwarks of this height would trap large quantities of water, overburdening so light a vessel. *Phantom* is 56 feet 2½ inches on deck, 51 feet 3 inches DWL, 13 feet 3 inches beam, and 4 feet 9 inches depth of hold, with 1,362 square feet in the working rig, 1,817 square feet total sail area, 17½ tons displacement to the DWL, and a cargo capacity of 13 tons.

The three-masted schooner is one of the handiest rigs ever devised for heavy weather, since, with a double-reefed main and the forestaysail, she happily keeps on moving with little for the crew to do other than catch up on their sleep. The limits of this rig seem to be from 50 to about 150 feet of deck length when used in conjunction with a bowsprit, or, if knockabout-rigged, 60 to 180 feet. Throughout this range, three-masters take less effort to sail than equivalent two-masted schooners. The *Sultana* is an example of one of the smaller vessels so rigged (Figure 25). Gaff-rigged

SAIL MAKERS NOTES:

1. SAILS TO BE TRIPLE STITCHED 13¾ VIVATEX OR 775 3 DACRON.
2. SAILS TO HAVE HAND SEWEN BOLT ROPES ALL AROUND - JIB WILL HAVE A ROPE LUFF
3. DIMENSIONS ARE PIN TO PIN FULLY STRETCHED - MAST HOOPS ARE 13" FOR FORE MAST & 11" FOR MAIN MAST - NOTE DWG OF TACK FITTING. LACING IN LIEU OF HOOPS - OWNERS OPTION
4. HEAD BOARDS ARE ¾" MAHOG MARINE PLY & ROPING TO BE CARRIED OVER & UNDER (ALL AROUND) SAIL IS X LACED TO HEAD BOARD WITH 20 THREADS OF SAIL TWINE.
5. REEFS ARE TO BE IN BANDS & NETTLES FITTED.
6. GROMMETS AT EACH SEAM ON FOOT WITH THIMBLE FITTED BEL BOLT ROPE
7. CLOTHS ARE 18" OR FALSE SEAMED IF VIVATEX
8. NO BATTENS IN ANY SAIL

SPAR MAKERS NOTES:

1. ALL SPARS TO BE 6061-T6 ALUMINUM ALLOY TO DIAS GIVEN & INDICATED
2. MAIN MAST - 43'8" LOA - LOWER 9'0" - 8" SCH 40 PIPE - UPPER 34'8" - GRAYBAR SECT 61-088 - ¼" WALL TAPERS 8" TO 4½" - OR MFG OF EQ SECT
3. MAIN BOOM - 27'0" LOA FROM AFT SIDE OF MAST (DEDUCT CLAPPER) 4" SCH 40 PIPE
4. FORE MAST - 51'0" LOA - LOWER 11'9" 10" SCH 40 PIPE - UPPER 39'3" - GRAYBAR SECT 61-122 - ¼" WALL - TAPERS 10" TO 6"
5. FORE BOOM - 24' LOA FROM AFT SIDE OF MAST (DEDUCT CLAPPER) 4" SCH 40 PIPE
6. BOW SPRIT - AS PER DWG #192-5
7. JIB CLUB 3" DIA OAK OR ASH
8. THUMBS (FOR SHROUDS) BLOCK FITS - MAST HEAD & OTHER FITTINGS AS PER MAKERS STD - BUT NOT LESS THAN ½" MAT. UNLESS NOTED OTHERWISE
9. AFT SIDE OF MAST IS STRAIGHT
10. FIT BOOMS WITH EYES FOR JACK LINE -
11. ETCH & PAINT ALL SPARS

EQUIPMENT:

1. 2-45# COR PLOW - BOWERS.
2. 20# DANFORTH - KEDGE
3. 45 FATHOMS ⅜" BBB GALV. CHAIN - EA BOWER - CORAL BOTTOMS
   5 FATHOMS ⅜" BBB GALV CHAIN + 50 FATH ¾ NYLON - EA BOWER - SAND BOTT.
4. LIFE JACKETS FOR EA. BERTH + 1 EXTRA - USCG OR BOT TYPE
5. MARINASPEC TRI-COLOR - ON FORE MAST TOP # M7001 - (MOTOR VESSELS REQ EXTRA LIGHTS)
6. ANCHOR LIGHT (OIL)
7. FOG HORN & FOG BELL
8. BILGE PUMP (MANUEL) 2" SUCTION & DISCHARGE (EDSON)
9. SESTRAL 5" SPHERICAL COMPASS FOR STEEL HULLS
10. 12' BOAT HOOK
11. 4½# SOUNDING LEAD & MARKED LINE (10 FATH)
12. 300,000 CP PORTABLE LIGHT
13. 6 MOORING LINES - ¾" NYLON - 8 FATHOMS. EA WITH A 20" EYE SPLICED IN ONE END & SEIZING ON OTHER END
14. AWNINGS & DODGERS & SAIL COVERS TO SUIT OWNER REQ
15. NAME & PORT OF HAIL ON STERN - NAME ON EA SIDE OF BOW & TRAIL BOARDS
16. 10' DINGHY ON DAVITS - SEE NOTE ON CONST. PLAN
17. 2 DRY CHEM. FIRE EXT (SAILING VESSEL) OR 3 IF MOTOR.
18. ANY & ALL OTHER EQUIPMENT REQ BY USCG OR FLAG OF REGISTER & REQ FOR SEA KEEPING ABILITY & SAFETY
19. LUNENBURG RACHET GYPSY WINDLASS - SIZE #2 FOR CHAIN - 2 WILDCATS +
20. " " " " " " #1 FOR ROPE - 2 WARPING HEADS.

RIGGERS NOTES:

1. STANDING RIGGING TO BE 7×7 GALV. IMPVD PLOW STEEL OR S. STEEL SHROUDS & JIB STAY ⅜" & TRIATIC & HEAD STAY 5/16"
2. ALL TERMINALS TO BE "NICOPRESS" SPLICING SLEEVES OVER THE PROPER SIZE THIMBLE WHEN REQ
3. EYES OVER MASTS TO BE WORMED, PARCELED & SERVED
4. SHROUDS ARE SINGLE ON MAIN MAST & ARE PASSED OVER AS PAIRS ON FORE MAST FROM ONE CONT WIRE & ARE MARRIED WITH A "NICOPRESS" SLEAVE
5. TURNBUCKLES ARE ⅝" JAW & JAW GALV. WITH LOCK NUTS
6. BOBSTAY IS ½" GALV CHAIN.
7. LOWER END OF JIB STAY IS ⅜" GALV CHAIN.
8. LAZY JACKS ARE 5/16" (3) STRAND DACRON. - SEE SKETCH - FOR JIB. FORE & MAIN CAN BE SET UP P&S
9. TOPPING LIFTS ARE ⅜" (3) STRAND DACRON.
10. HALYARDS & SHEETS TO BE ½" (3) STRAND DACRON.
11. JIB DOWN HAUL TO BE 5/16" DACRON - 3 STRAND
12. BLOCKS TO BE GIBBS OR EQUAL
13. HALYARDS, LAZY JACKS BELAY ON PIN RAILS P/S. AT RAIL CAP LEVEL · FORE MAST MAY ALSO HAVE A PIN RAIL IN SHROUDS TO HANG COILED LINES. - CLEAT ON STB MAIN SHROUD FOR CB. LANYARD. - LIFTS BELAY ON BOOMS - NOT SHOWN IS REEF PENDANTS - USE STD SCHOONER DETAIL.
14. JIB SHEET IS RUN AFT VIA FAIR LEADERS ATTACHED TO RAIL CAP SIDES FROM SNATCH BLOCK - STBD ONLY.

Figure 24. *On this and the three following pages:* Phantom, *a 13-tons-capacity cargo schooner, Colvin Design No. 192.*

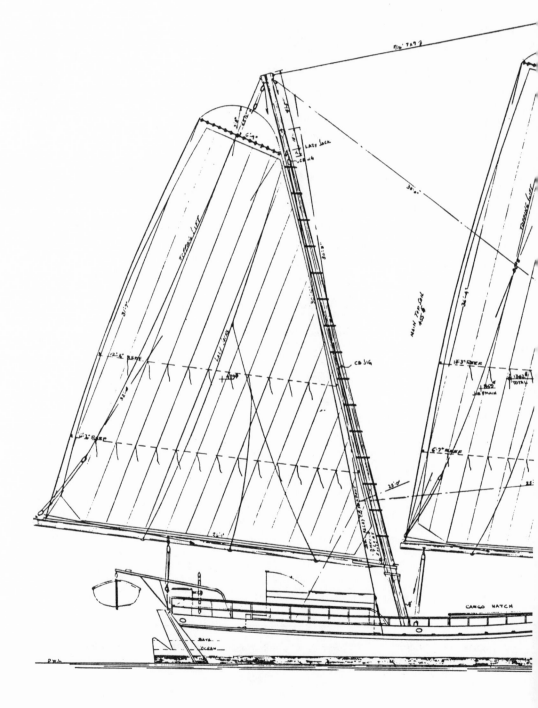

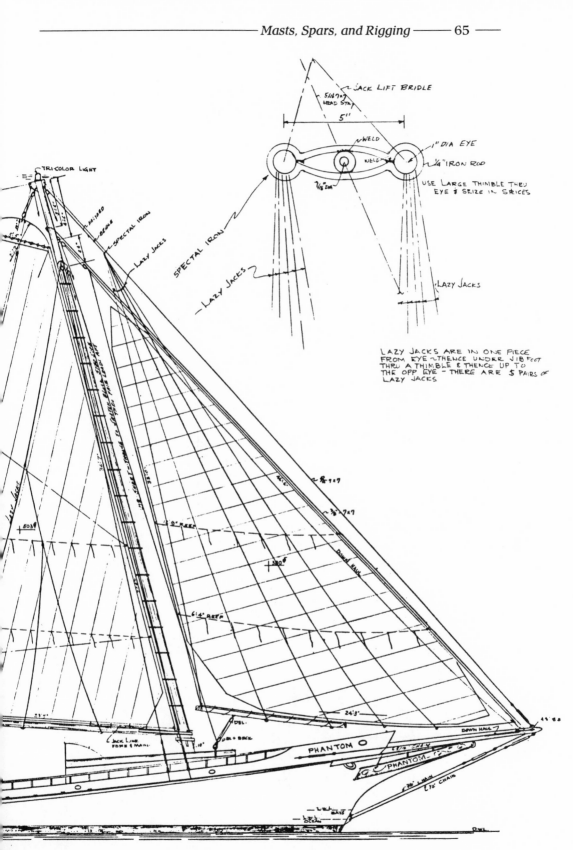

JACK LIFT BRIDLE

5/16 7x7 HEAD STAY

5"

WELD

1" DIA EYE

WELD

1/4" IRON ROD

USE LARGE THIMBLE THRU EYE & SEIZE IN SPLICES

7/8" DIA

SPECTAL IRON

LAZY JACKS

LAZY JACKS

LAZY JACKS ARE IN ONE PIECE FROM EYE — THENCE UNDER JIB FOOT THRU A THIMBLE & THENCE UP TO THE OPP EYE — THERE ARE 5 PAIRS OF LAZY JACKS

TRI-COLOR LIGHT

HALYARD

SPECTAL IRON

LAZY JACKS

12" REEF

6' 4" REEF

DOWN HAUL

24' 3"

DOWN HAUL

PHANTOM

PHANTOM

EYE CHAIN

D.W.L.

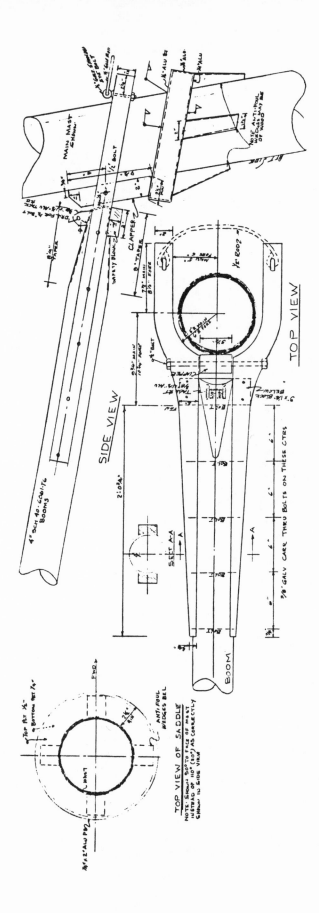

ITEM	NAME	SIZE	TYPE	CAT NO.	LOCATION
1	MIZ. GAFF TOPSL HALYARD	3½"	SING	23	PADEYE ON MIZ. TOP MAST
2	MIZ GAFF TOPSL OUT HAL	3½"	SING	31	PADEYE ON MIZ GAFF
3	MIZ. HALYARD	4"	SING + BECK	31	PADEYE ON MIZ MAST
4	MIZ HALYARD	4"	SING	44	GAFF BRIDLE
5	MIZ HALYARD	4"	SING	23	PADEYE ON MIZ. MAST
6	MIZ HALYARD	4"	SING	31	STRAP ON GAFF JAWS
7&8	MIZ LAZY JACKS	3½"	SING	31	1-PORT 1STBD PADEYES MIZ MAST
9-10&11	MIZ LAZY JACK BRIDLE	3"	SING	35	2-PORT, 2 STBD
13	MIZ SHEET	4"	DOUB + BECK	23	MIZ BOOM BALE
14	MIZ SHEET	4"	DOUBLE	25	MIZ RAIL TRAVELER
15	MIZ SHEET FAIRLEAD	4"	SING	54	WHEEL BOX
16	MIZ PREVENTER	3"	SING + BECK	24	BRIDLE ON MIZ BOOM
17	MIZ PREVENTER	3"	SING	30	TO EYE ON RAIL CAP P/S
18	MAIN GAFF TOPSL HAL	3½"	SING	23	PADEYE MAIN TOP MAST
19	MAIN GAFF TOPSL OUT HAUL	3½"	SING	31	PADEYE MAIN GAFF
20	MAIN HALYARD	4"	SING + BECK	31	PADEYE MAIN MAST
21	MAIN HALYARD	4"	SING	44	GAFF BRIDLE
22	MAIN HALYARD	4"	SING	23	PADEYE ON MAIN MAST
23	MAIN HALYARD	4"	SING	31	STRAP ON GAFF JAWS
24-25	MAIN LAZY JACKS	3½"	SING	31	1-PORT 1-STBD PADEYES MAIN MAST
26-27	MAIN LAZY JACK BRIDLE	3"	SING	35	1-PORT 1-STBD
28	MAIN SHEET	4"	DOUB + BECK	23	MAIN BOOM BALE
29	MAIN SHEET	4"	DOUB	25	MAIN TRAVELER TOP OF AFT CABIN
30	MAIN SHEET FAIRLEAD	4"	SING	54	ON ℄ TOP OF AFT CABIN
31	MAIN PREVENTER	3"	SING + BECK	24	BRIDLE ON MAIN BOOM
32	MAIN PREVENTER	3"	SING	80	TO EYE ON RAIL CAP P/S
33	FORE GAFF TOPSL HAL	3½"	SING	23	PADEYE FORE TOP MAST
34	FOR GAFF TOPSL OUT HAUL	3½"	SING	31	PADEYE FORE GAFF
35	FORE HALYARD	4"	SING + BECK	31	PADEYE ON FORE MAST
36	FORE HALYARD	4"	SING	44	GAFF BRIDLE
37	FORE HALYARD	4"	SING	23	PADEYE ON FORE MAST
38	FORE HALYARD	4"	SING	31	STRAP ON GAFF JAWS
39-40	FORE LAZY JACKS	3½"	SING	31	1-PORT 1STBD PAD EYES FORE MAST
41-42	FORE LAZY JACK BRIDLE	3"	SING	35	1-PORT 1-STBD
43	FORE SHEET	4"	DBL + BECK	23	FORE BOOM BALE
44	FORE SHEET	4"	DOUBLE	25	FORE TRAVELER ON FWD CABIN
45	FORE SHEET FAIRLEAD	4"	SING	54	℄ FORE CABIN TOP
46	FORE PREVENTER	3"	SING + BECK	24	BRIDLE ON FORE BOOM
47	FORE PREVENTER	3"	SING	30	TO EYE ON RAIL CAP P/S.
48-49	SQ'SL LIFTS	4"	SING	31	1-P 1-S - FORE MAST CAP ℄
50-51	SQ'L BRACES	4"	SING + BECK	31	1-P 1-S PADEYE ON MAIN MAST P/S
52-53	SQ'SL BRACES	4"	SING	31	1-P-1-S ON END OF YARD P/S
54-55	SQ'SL SHEETS	4"	SING + BECK	26	1-P 1-S CLEWS OF SAIL P/S
56-57	SQ'SL SHEETS	4"	SING	30	1-P 1-S PADEYES ON RAIL P/S
58-59	SQ'SL SHEETS FAIRLEAD	4"	SING	30	1-P. 1-S PAD EYE ON RAIL P/S
60-61	BUNT LINE FALLS	3"	SING	26	1-P-1-S 12" OFF ℄ ON YARD. P/S.
62	STAYSAIL HALYARD	4"	SING + BECK	31	PADEYE ON FORE MAST
63	STAYSAIL HALYARD	4"	SING	31	ON STAYSAIL
64	STAYSAIL SHEET	4"	SING + BECK	23	BALE ON STAY'L BOOM
65	STAYSAIL SHEET	4"	SING	25	FREE STS'L TRAVELER FORE DECK
66	STAYS'L SHEET FAIRLEAD	4"	SING	54	ON BREAST FITT AT STEM
67	JIB HALYARD	4"	SING + BECK	31	PAD EYE ON FORE MAST
68	JIB HALYARD	4"	SING	31	ON JIB
69-70	JIB SHEET	4"	SING + BECK	25	1-P 1-S ON BRIDLE JIB CLEW
71-72	JIB SHEET	4"	SING	31	1-P 1-S PADEYE MAIN DK. P/S
73-74	JIB SHEE FAIRLEAD	4"	SING	31	1-P 1-S PADEYE MAIN DK P/S.
75	JIB DOWN HAUL	3"	SING	23	STBD SIDE END OF BOWSPRIT
76	SING SPANISH BURTON	3½"	SING	35	TO WHIP ON PADEYE BEL YARD
77	SING SPANISH BURTON	3½"	SING	35	TO BLOCK # 76 RUNNING WHIP
78-79	DAVIT FALLS	4"	DBL + BECK	46	TO DAVITS P/S.
80-81	DAVIT FALLS	4"	DBL	32	NO BECK - TO DINGHY P/S.
82	HANDY BILLY	4"	DBL	33	NO BECK GEN PURPOSE
83	HANDY BILLY	4"	SING + BECK	32	GEN PURPOSE
	SPARES				TO SUIT OWNER REQ

Figure 25. *The block list, Sail Plan (page 68) and Sailmaker and Spar-maker's Notes (page 69) for* Sultana, *a 53-foot three-masted auxiliary schooner, Colvin Design No. 122.*

and pole-masted, she sets gaff topsails and a squaresail. The staysails were omitted originally but have since been added to some of her sisterships, because the absence of a springstay makes these sails easy to tack. Note that the sails are of similar size: experience has proven this most satisfactory arrangement, for when the usual large mizzen is fitted, it seems to take charge when reaching and running, causing the vessel to be hardmouthed and to sheer. In some instances, it is possible to design the

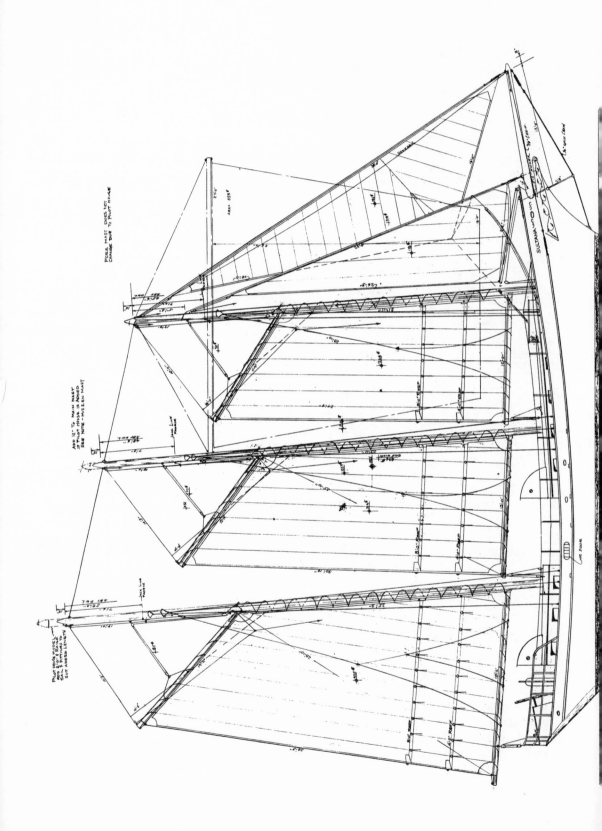

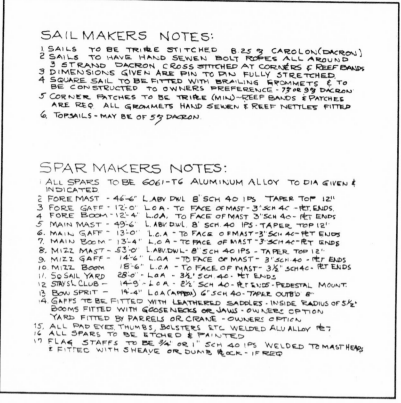

SAILMAKERS NOTES:
1. SAILS TO BE TRIPLE STITCHED 8.25 % CAROLON(DACRON)
2. SAILS TO HAVE HAND SEWEN BOLT ROPES ALL AROUND
   3 STRAND DACRON CROSS STITCHED AT CORNERS & REEF BANDS
3. DIMENSIONS GIVEN ARE PIN TO PIN FULLY STRETCHED.
4. SQUARE SAIL TO BE FITTED WITH BRAILING GROMMETS & TO
   BE CONSTRUCTED TO OWNERS PREFERENCE - 7% OR 9% DACRON
5. CORNER PATCHES TO BE TRIPLE (MIN)~REEF BANDS & PATCHES
   ARE REQ. ALL GROMMETS HAND SEWEN & REEF NETTLES FITTED
6. TOPSAILS - MAY BE OF 5% DACRON.

SPAR MAKERS NOTES:
1. ALL SPARS TO BE 6061-T6 ALUMINUM ALLOY TO DIA GIVEN &
   INDICATED.
2. FORE MAST - 46'-6" L.ABV DWL 8" SCH 40 IPS TAPER TOP 12"
3. FORE GAFF - 12'-0" LOA - TO FACE OF MAST - 3" SCH 4C - FLT. ENDS.
4. FORE BOOM - 12'-4" L.OA, TO FACE OF MAST 3" SCH 4c - FLT ENDS
5. MAIN MAST - 49'-6" L.ABV DWL. 8" SCH.40 IPS - TAPER TOP 12"
6. MAIN GAFF - 13'-0" L.O.A - TO FACE O F MAST - 3" SCH 4C- FLT ENDS
7. MAIN BOOM - 13'-4" L.O.A - TO FACE OF MAST - 3" SCH 4C - FLT ENDS
8. MIZZ MAST - 53'-0" L.ABV.DWL - 8" SCH 40 IPS - TAPER TOP 12"
9. MIZZ GAFF - 14'-6" L.OA - TO FACE OF MAST - 3" SCH.40 - FLT ENDS
10. MIZZ BOOM - 18'-6" L.CA - TO FACE OF MAST - 3½" SCH4C- FLT ENDS
11. SQ SAIL YARD - 28'-0" - LOA - 3½" SCH. 4C- FLT ENDS
12. STAYS'L CLUB - 14'-9" - L O A - 2½" SCH 40- FLT ENDS - PEDESTAL MOUNT.
13. BOW SPRIT - 14'-4" LOA (APPROX) 6" SCH 40- TAPER OUTB'D 8"
14. GAFFS TO BE FITTED WITH LEATHERED SADDLES - INSIDE RADIUS OF 5½"
    BOOMS FITTED WITH GOOSENECKS OR JAWS - OWNERS OPTION
    YARD FITTED BY PARRELS OR CRANE - OWNERS OPTION
15. ALL PAD EYES, THUMBS, BOLSTERS ETC WELDED ALU ALLOY FLT
16. ALL SPARS TO BE ETCHED & PAINTED
17. FLAG STAFFS TO BE ¾' OR 1" SCH 40 IPS WELDED TO MASTHEADS
    & FITTED WITH SHEAVE OR DUMB BLOCK - IF REQ

rig so that the fore, main, and mizzen sails are identical and interchangeable. The block list calls for 83 blocks, not including spares; however, *Sultana* uses no winches. She was fitted out for charter work, but several of her sisters are engaged in general freighting, and another one is a floating machine shop. Her dimensions are 52 feet 10 inches on deck, 43 feet 4 inches DWL, 15 feet 6 inches beam, 6 feet 3 inches depth of hold, and 1,325 square feet of sail area in the five lowers, with a total sail area of 2,116 square feet and 31 tons displacement at the DWL. When carrying cargo, she loads an additional 20 tons.

When someone mentions a three-masted schooner, the image immediately comes to mind of three gaff-rigged sails, several headsails, gaff topsails, and topmast staysails on a heavy displacement hull. *Sultana* certainly fits this description. On the other hand, none of these images applies at all to the three-masted, jibheaded schooner, *Irregardless* (Figure 26), 50 percent longer than *Sultana* with one-fifth less displacement. Her light draft is but 2 feet 9½ inches, and she sails without ballast. This vessel has no engine, but carries her motorized yawlboat in davits over the stern, which allows her to be pushed in calms, up rivers against the current, and to maneuver in berthing. Due to her light displacement and shallow draft, solid bulwarks would cause a stability problem in a seaway, so she has open rails. She carries two workboats on deck. *Irregardless* sails with a crew of two but can easily be

Figure 26. *A three-masted sharpie cargo schooner (Colvin Design No. 197) having only 2 feet 9½ inches of light draft (11 feet 7 inches with her centerboard down).*

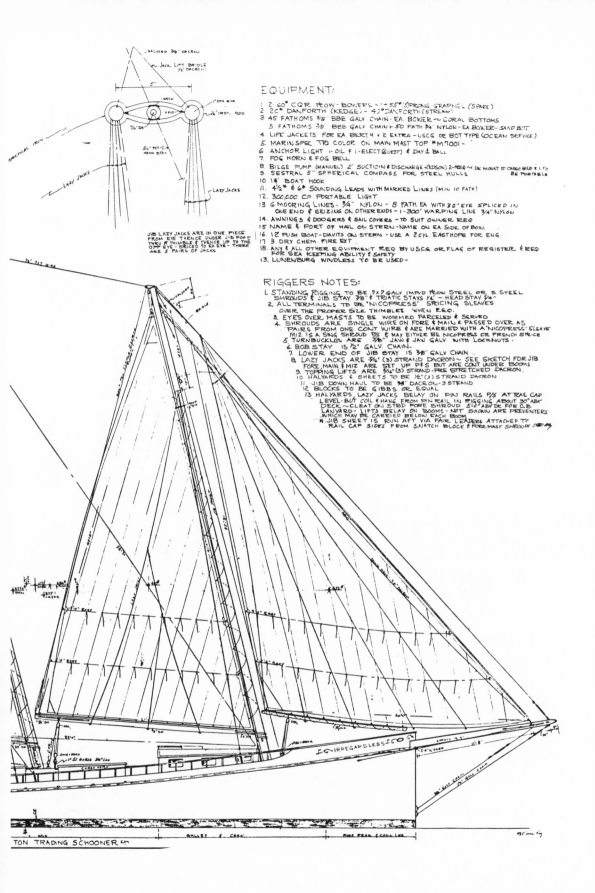

EQUIPMENT:

1. 2·60# CQR PLOW-BOWERS ~ 1-55# (5)PRONG GRAPNEL (SPARE)
2. 2C# DANFORTH (KEDGE) ~ 4J#DANFORTH (STREAM)
3. 45 FATHOMS 3/8" BBB GALV CHAIN-EA. BOWER ~ CORAL BOTTOMS
   5 FATHOMS 3/8" BBB GALV CHAIN + 50 FATH 3/4" NYLON ~ EA BOWER- SAND BOTT
4. LIFE JACKETS FOR EA. BERTH + 2 EXTRA ~ USCG OR BOT TYPE (OCEAN SERVICE)
5. MARIN SPEC TRI COLOR ON MAIN MAST TOP # M 7001 ~
6. ANCHOR LIGHT 1- OIL & 1- ELECT (GUEST) & DAY & BALL
7. FOG HORN & FOG BELL
8. BILGE PUMP (MANUEL) 2" SUCTION & DISCHARGE ~(EDSON) 2-REQ ~1 DK MOUNT TO CARGO HOLD & 1 TO
9. SESTRAL 5" SPHERICAL COMPASS FOR STEEL HULLS   BE PORTABLE
10. 14' BOAT HOOK
11. 4 1/2# & 6# SOUNDING LEADS WITH MARKED LINES (MIN 10 FATH)
12. 300,000 CP PORTABLE LIGHT
13. 6 MOORING LINES - 3/4" NYLON - 8 FATH. EA WITH 30" EYE SPLICED IN
    ONE END & SEIZING ON OTHER ENDS - 1-300' WARPING LINE 3/4" NYLON
14. AWNINGS & DODGERS & SAIL COVERS - TO SUIT OWNER REQ
15. NAME & PORT OF HAIL ON STERN - NAME ON EA. SIDE OF BOW
16. 12 PUSH BOAT - DAVITS ON STERN - USE A 2 CYL. EASTHOPE FOR ENG
17. 3. DRY CHEM FIRE EXT.
18. ANY & ALL OTHER EQUIPMENT REQ BY USCG. OR FLAG OF REGISTER & REQ
    FOR SEA KEEPING ABILITY & SAFETY
19. LUNENBURG WINDLESS TO BE USED~

RIGGERS NOTES:

1. STANDING RIGGING TO BE 7x7 GALV IMPD PLOW STEEL OR S. STEEL
   SHROUDS & JIB STAY 3/8" & TRIATIC STAYS 1/4" ~ HEAD STAY 5/16"
2. ALL TERMINALS TO BE "NICOPRESS" SPLICING SLEEVES
   OVER THE PROPER SIZE THIMBLES WHEN REQ.
3. EYES OVER MASTS TO BE WORMED PARCELED & SERVED
4. SHROUDS ARE SINGLE WIRE ON FORE & MAIN & PASSED OVER AS
   PAIRS FROM ONE CONT WIRE & ARE MARRIED WITH A "NICOPRESS" SLEAVE
   MIZ IS A SNG SHROUD P/S & MAY EITHER BE NICOPRESS OR FRENCH SPLICE
5. TURNBUCKLES ARE 3/8" JAW & JAW GALV WITH LOCKNUTS.
6. BOB STAY IS 1/2" GALV. CHAIN·
7. LOWER END OF JIB STAY IS 3/8" GALV CHAIN
8. LAZY JACKS ARE 5/16" (3) STRAND DACRON ~ SEE SKETCH FOR JIB.
   FORE, MAIN & MIZ ARE SET UP P/S BUT ARE CONT. UNDER BOOMS
9. TOPPING LIFTS ARE 5/16" (3) STRAND-PRE STRETCHED DACRON.
10. HALYARDS & SHEETS TO BE 1/2" (3) STRAND DACRON-3 STRAND
11. JIB DOWN HAUL TO BE 3/8" DACRON
12. BLOCKS TO BE GIBBS OR EQUAL
13. HALYARDS LAZY JACKS BELAY ON PIN RAILS P/S AT RAIL CAP
    LEVEL - BUT COIL & HANG FROM PIN RAIL IN RIGGING ABOUT 30" ABV
    DECK ~ CLEAT ON STBD FORE SHROUD 54" ABV DK FOR C.B.
    LANYARD ~ LIFTS BELAY ON BOOMS - NOT SHOWN ARE PREVENTERS
    WHICH MAY BE CARRIED BELOW EACH BOOM
14. JIB SHEET IS RUN AFT VIA FAIR LEADERS ATTACHED TO
    RAIL CAP SIDES FROM SNATCH BLOCK & FOREMAST SHROUDS STBD ONLY

HALYARD 3/8" DACRON
JACK LIFT BRIDLE 1/2" DACRON
5'
WELD
1" DIA EYE
3/16" IRON. ROD
SPHERICAL IRON
WELD
3/16" DIA
5/16" PATT ~
HEAD STAY
LAZY JACKS
LAZY JACKS

JIB LAZY JACKS ARE IN ONE PIECE
FROM EYE THENCE UNDER JIB FOOT
THRU A THIMBLE & THENCE UP TO THE
OTH EYE - SPLICED TO EA EYE - THERE
ARE 5 PAIRS OF JACKS

IRREGARDLESS & CO

TON TRADING SCHOONER
HOLD   GALLEY & CABIN   FORE PEAK & CHAIN LKR

singlehanded and has sparse accommodations for six passengers. Her cargo capacity is 35 tons maximum; the normal load is 27 tons of general freight or building materials in the hold, along with a deck cargo of lumber. The mainmast steps off the centerline to port, with the centerboard sited off the centerline to starboard. She measures 78 feet 3 inches on deck, 69 feet 1 inch DWL, 15 feet 9 inches beam, 5 feet 9 inches depth of hold, and 1,695 square feet of sail area in the four lowers, with a total sail area of 2,774 square feet and a displacement of 25 tons to the DWL.

There is no reason to accept any modern trend as the only means of achieving an efficient rig. Most modern rigs require numerous different sails to replace ones already in use on a spar or stay in order to make them efficient, but when commercial vessels have light sails, these are always used *in addition to* their working sails. In all instances, the sails are cut up-and-down in the old-fashioned way (just recently invented, if you believe the advertising media), and these sails have no battens.

MISCELLANEOUS DETAILS

Figure 27 shows the details of gaff and boom jaws for two three-masted schooners. The same dimensions are also used on the Pinky *Sea Wolf*, illustrating this book. Jaws make better shipmates than goosenecks and are less trouble for the builder to make. Whether on a yacht or commercial vessel, the use of jaws allows reefing the boom up for better visibility and to clear any deck cargoes. In port, one can raise the whole sail after furling and gasketing, to permit full headroom under the booms and the use of the booms as ridgepoles for an awning. Jaws impose less strain on the mast. On commercial vessels, the gaffs double as cargo booms. Figure 28 shows several ways in which vessels fitted with jaws can utilize them, and how the cargo is worked with a gaff. The figure also shows calculations for a 1,000-pound lift using the gaff. This is about the maximum lift for the smaller vessels, and most masters keep the lift to 500 pounds because the stowage of heavy pieces of freight in the hold is difficult at best, unless they remain in the hatchway. Deck loads such as automobiles and tractors are driven onto the deck rather than lifted on.

Both *Phantom* and *Irregardless* use a jackline on the boom, a ¼-inch 7 x 7 wire rove through eyelets screwed into the boom on approximately 18- to 21-inch centers, which evens the tension on the foot of the sail and at the same time makes it easier to pass the reef nettles under the boltrope. The preferred sail material on commercial vessels is water-repellent, mildew-resistant cotton, especially in the tropics, since the sails then need not be covered, last longer than synthetic cloth, and are easier to repair. Sail covers are sometimes used during lengthy stays in port or when a dusty cargo is to be loaded or discharged. Even then, it is best to air out the sails frequently. The daily covering of sails is a luxury reserved for yachts.

The attachment of blocks for the peak and other halyards can be done with strops passed around the mast and held up by welded thumbs. Welded eyes on metal masts or eyebands on wooden spars will also serve to attach blocks. Eyes are the least expensive alternative and are therefore used most often. A bolster will help keep the

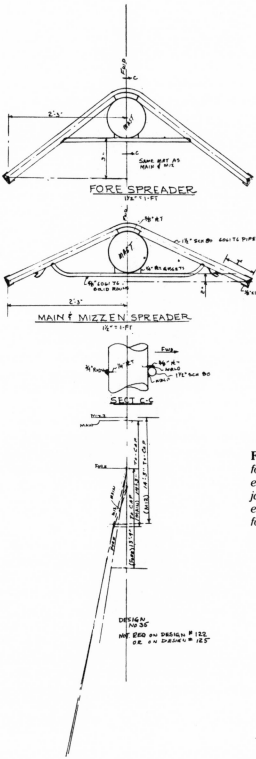

FORE SPREADER
1½" = 1-FT

MAIN & MIZZEN SPREADER
1½" = 1-FT

SECT C-C

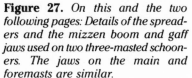

DESIGN
No 35
NOT REQ ON DESIGN #122
OR ON DESIGN #125

Figure 27. *On this and the two following pages: Details of the spreaders and the mizzen boom and gaff jaws used on two three-masted schooners. The jaws on the main and foremasts are similar.*

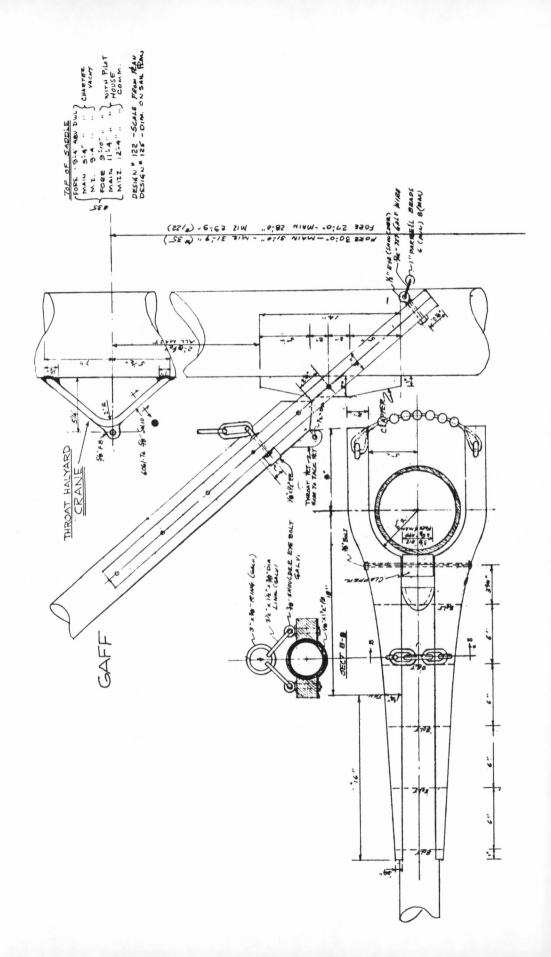

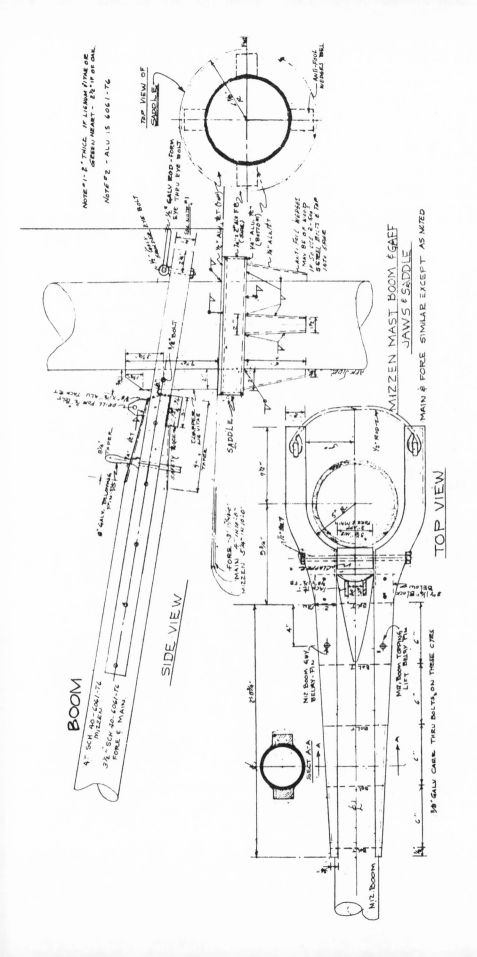

MIZZEN MAST BOOM & GAFF
JAWS & SADDLE

MAIN & FORE SIMILAR EXCEPT AS NOTED

TOP VIEW

SIDE VIEW

BOOM

4" SCH 40 - 6061-T6
MIZZEN

3½" SCH 40 - 6061-T6
FORE & MAIN

TOP VIEW OF
SADDLE

SECT A-A

NOTE #1 - 2" THICK IF LIGNUM VITAE OR
GREEN HEART - 2½" IF OF OAK.

NOTE #2 - ALU IS 6061-T6

SADDLE

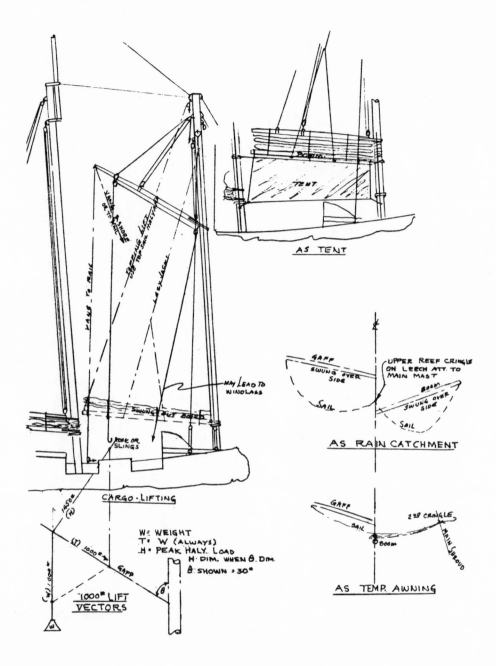

Figure 28. *Some uses for a gaff on a vessel having one.*

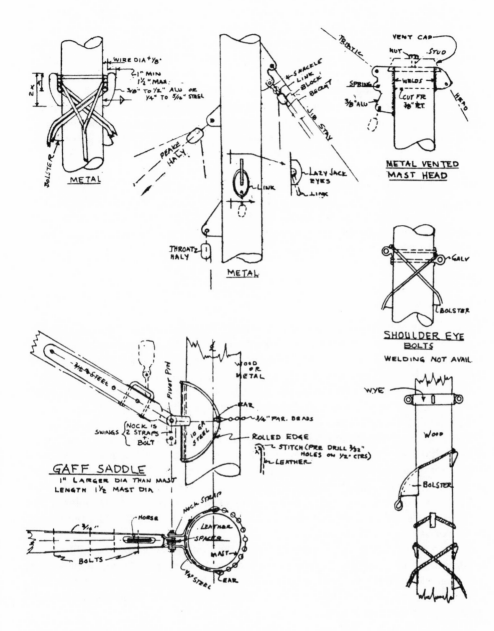

Figure 29. *Details of rigging attachments on masts, and a drawing of a gaff saddle* (bottom left) *that can replace jaws.*

throat halyard on smaller vessels clear of the mast if strops are used; however, a crane is better. Figure 29 shows sketches of the various details for securing the rigging to the mast. This figure also shows the details of a gaff saddle used in lieu of jaws on yachts and some commercial vessels. Saddles can be fabricated by the builder, after which they should be galvanized and covered with leather. The saddles have a greater bearing surface on the mast and, on wooden spars, eliminate the sheathing that is required when jaws are used. Improper design—either too large a diameter or too short a length—will cause the saddles to capsize when raising and sometimes when lowering the sail, but a point in their favor is that the nock of the sail need not be cut back as much as when jaws are used.

Triatic and springstays, when used, should be cut to the exact length and set up with shackles only. The use of turnbuckles aloft adds extra weight and expense and only invites trouble since, when fouled, turnbuckles have a tendency to bend in the threaded portion and thus become weakened. Also, check nuts have a tendency to work loose, slackening the stay; because of their location, the check nuts are also a nuisance to get to, let alone adjust.

SPARS

Solid Spars

A builder should have a good working knowledge of how to lay out and fabricate solid wooden spars, since they are still used on steel vessels; also, he will have to fabricate all of the spar hardware formerly associated with wooden vessels. It is true that metal spars have become popular; however, this has more to do with the availability of wood of sparmaking quality than with the superiority of metal. In any case, when one is constructing a hollow metal spar, the same principles as for wood will apply, especially if the spar is tapered.

The woods primarily used in sparmaking are Douglas fir, pine, and spruce. Other woods that have been used include oak, teak, and the mahoganies. The most important qualities to look for are straight, dense grain free of knots, shakes, pitch pockets, and sapwood. The weight of the wood is secondary to its glueability and workability. The most popular woods for gluing are Sitka spruce and Douglas fir. For solid spars, the best woods would be Sitka spruce, Douglas fir, rosemary pine, white pine, Norway pine and Riga fir.

In large, solid masts, the presence of small tight knots is not objectionable, but in small spars—booms, gaffs, and topmasts—the material should be free of all defects. The heartwood of the tree is the best wood for masts. Whether the material being worked is square or round, the center of the heart should lie as near as possible to the center of all the sections.

To make a spar, one first sets up the timber on several sawhorses and, if it is round (the actual tree), dubs off a small portion on the underside so that the timber will not roll. If the unfinished diameter approximates the actual finished size, curved wedges must be fitted at each horse to prevent turning during the initial shaping. A plumb line through the heart is established at each end of the timber and marked on the top. Between these two marks, a chalkline is stretched and plucked, leaving the centerline as a chalked line.

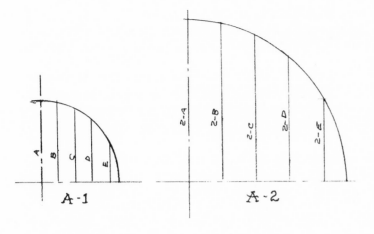

Figure 30. *A mast tapered according to a diagram (A-1) appears as a straight line from a distance. When the afterside of a mast is left straight to accept a sail track, the forward side is given twice the normal taper, as derived from A-2. Sketches B and C show the two types of taper as viewed from one side.*

Some builders use a straight taper—that is, straight from the partners to the head—which always imparts the illusion that the spar is concave. When the spar is tapered according to a diagram, as shown in Figure 30, it will appear as a straight line from a distance. If, for example, a square timber is used, the appropriate tapers are derived from a diagram, laid down on the timber, and trimmed. These trimmed faces become the port and starboard sides of the spar. The afterside of the mast must be straight only if sail track is to be used, in which case the forward face has double the side taper. Figure 30 shows a mast on which sail track is used. For all other spars, the side taper is the same as the fore-and-aft taper, and all sides of the spar diminish equally. Straight taper is used from the boom to the heel (bury), which is about 80 percent of the mast diameter at the boom. This taper is best for wedging the partners. Booms have their top edges straight, with all tapering done on the sides and bottom. Gaffs have their bottoms straight and all taper on the top and sides. Thus, the sparmaker is always working from a straight edge. Bowsprits may be straight-tapered (spike), circular-tapered (spar), top-tapered (modified hog), or hogged and tapered (in which case the top and bottom curve downward in profile). Bowsprits may also take the form of planks, or be A- or U-shaped. Figure 31 shows

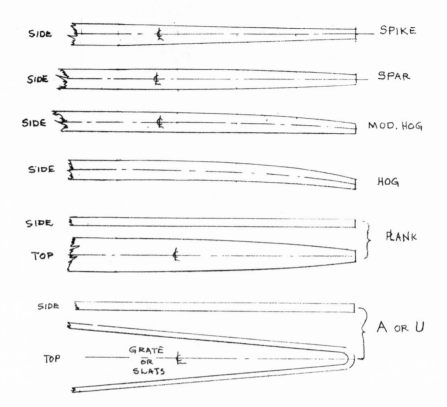

Above: Figure 31. *Common bowsprit configurations.* **Below: Figure 32.** *A uniformly tapered yard.*

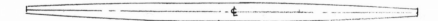

the various bowsprit configurations. Yards have a uniform taper on all sides (Figure 32), and the layout is done from the centerline of the timber. Masts are made as the tree grows, with the butt end always down.

Once the centerline is established, it is customary first to hew or cut the vertical sides of the timber so that each face has the same offset from the centerline, forming the sides of the roughed-out spar. In the case of a mast to be fitted with sail track, however, the side that is to face aft, being a straight edge, is the first edge made and trued. The forward side will then be trimmed to have twice the taper to be used on the port and starboard sides. When this has been accomplished, the spar is turned 90

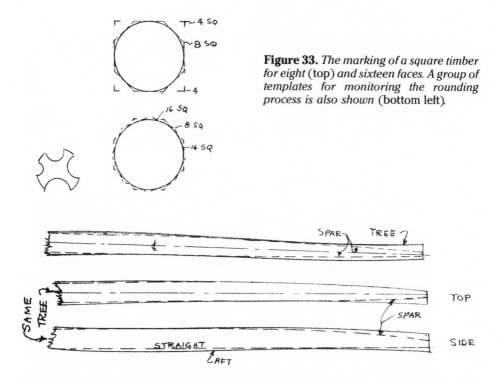

Figure 33. *The marking of a square timber for eight* (top) *and sixteen faces. A group of templates for monitoring the rounding process is also shown* (bottom left).

Figure 34. Top: *A mast for a gaff rig being worked out of a crooked tree.* Bottom: *Two views of a spar being made from a crooked tree for a jibheaded rig.*

degrees to expose the other two surfaces, with the afterside facing up. The centerline is again established, scribed, and marked for tapering. The side tapers are cut, leaving a square timber that may now be rounded in stages, as shown in Figure 33. First it is marked and cut for eight equal faces, then marked and cut again for 16 faces; large spars will require 32 faces. The rounding is then commenced, the sparmaker working mostly by feel and eye, which is extremely accurate. Some sparmakers use a series of templates that are exactly half-rounds as a check; however, perfection of this sort is not required except in the way of bands or other ironwork that must fay all around.

A tree intended to be made into a mast may, at the butt end, have an unfinished diameter almost the correct diameter needed at the boom. When track will be used, only one face of the tree need be worked to straighten out any natural or double bend inherent in the tree. Figure 34 indicates a mast for a gaff rig being worked out of a crooked tree. The centerline is scribed and the layout of tapers marked. Note that the centerline must skew away from the actual center of the tree. Figure 34 also shows a spar being made for a jibheaded rig, using the same tree in both views. The straight after edge is scribed first, then the forward taper is worked in as shown. The log is then turned, and the side taper is worked from a centerline.

The correct amount of taper for a gaff-rigged wooden mast will be at least 1 percent but seldom more than 10 percent of the diameter at the boom, measured from the boom to where the jaws of the gaff will ride. Most builders reckon that a taper of at least 1 inch is the minimum to keep the hoops from jamming, and so increase the diameter of the mast at the boom by the necessary amount if the resulting upper diameter would otherwise be too small. On larger masts, the hoops must also have a forward trip line to prevent jamming. With lacing or parrell beads, jamming seldom occurs. When hoops or lacing is used, the sailmaker can easily adjust the cut of the luff to match the spar. Using a straight after face on the mast requires more cutting than using a round taper.

Solid masts for use with jibheaded sails require a different application of the taper diagram to utilize sail track. All one need do is to keep the afterside of the mast straight, confining all the taper to the forward and athwartship sides; for the fore-and-aft layout, the diagram is redrawn as shown in the center sketch in Figure 30, or the dimensions of the lefthand sketch are doubled, using the afterside of the mast as the centerline. Weight has always been of concern to designers and builders, but it is not as critical as one is led to believe. When the mast becomes very slender for its length, more rigging is necessary, and this additional rigging not only may be heavier but will create more windage than a slightly larger mast section would.

Hollow Wooden and Metal Spars

The early jibheaded rigs (leg-of-mutton) had low aspect ratios, and the mast length really amounted to the length of a lower mast plus a topmast. At approximately the same location where the crosstrees would have been on a mast fitted with a topmast, a spreader was added, and the mast continued as one pole to its top. The shrouds below performed the same function as they always had; the spreader just replaced the crosstrees. However, the upper shroud continued on to the deck rather than terminating at the crosstrees (attached to futtock shrouds). As the aspect ratios increased, the pole was continued upward, another set of spreaders was inserted where the topgallant mast would have started, and a shroud from the masthead was passed through these spreaders, as well as through the lower spreaders, and thence to the deck. This is about what one sees today. None of this was possible before the invention of mast track. Prior to that time, the sail was set flying or hanked to a jackstay on any mast in excess of the shroud height.

To lighten spars, 19th-century builders tried many methods of hollowing them: boring a hole, using long augers from each end of the spar and meeting in the middle, which weakened the top of the mast by removing too much material; using staves (Figure 35) banded with iron at frequent intervals, which usually resulted in masts heavier than the solid ones; shaping the masts, cutting them in half lengthwise, gouging them out to a uniform wall thickness, then using only a few iron bands to bind them back together, which produced masts that failed under compression loading, since the glues then available were neither really waterproof nor anywhere near as strong as the wood; and building the spars of iron and steel riveted together (before electric welding), which produced spars that buckled because the material

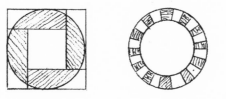

Figure 35. Right: *Cross-sectional diagram of a 19th-century mast construction that used staves banded with iron at frequent intervals.* Left: *An alternative to staving that is still used.*

was often too thin. When reliable waterproof glues became available, staving had a short revival, but it remained expensive. Figure 35 shows an alternative to staving often used to this day.

When welding of thin materials became possible (the coated electrode was invented in 1902), steel spars could be used, and were in use on the smaller vessels by 1920.

The advent of a relatively cheap reduction process in 1886 reduced the price of aluminum from $11.33 per pound to 23 cents per pound by 1900. The development of high-strength alloying aluminum came in the 1920s, and by World War II, welding techniques had steadily advanced; therefore, in reality, while the material is certainly not new, the ability to fabricate intricate aluminum structures is recent. In small steel vessels its primary use is for masts and other spars.

Most people take for granted the modern extrusions, welding, and spar laminations. These conveniences owe their existence to the development of economical means of extrusion, electric welding, the necessary filler wires, and waterproof adhesives of great strength. Most of this development has occurred within my own lifetime, so I find it quite simple to make comparisons between the way things used to be and the way they are done now. One could wish that progress always meant the improvement of a product, but it often means the discontinuation of a good product and the substitution of another, less expensive and inferior. The old ways are not always the best ways, but they are proven ways, and this is especially true of sparmaking and rigging.

On a jibheaded rig using hollow wooden spars, the rigging would be secured to the mast via tangs. Figure 36 shows those required for two 45-foot ketch designs. The location of the screws is most important in wooden spars, since the basic loading of screws is drag. In this example, the maximum loading per screw is 124 pounds. The outward pull has to be handled with bolts, but at the same time, the bolts are subject to drag loads, just as the screws are. Hollow bolts are made of pipe in order to save weight and increase the resistance to drag loads. The calculated loading is 2,100 pounds for a ¾-inch diameter. Solid bolts are used in several locations that require more strength in tension than in drag. The calculated strength of each lower tang assembly for the mainmast is 6,812 pounds. All material used is ³⁄₃₂-inch-thick mild steel, welded and drilled prior to galvanizing. Galvanized tang assemblies last longer than those made of stainless steel, since they are not subject to stress corrosion.

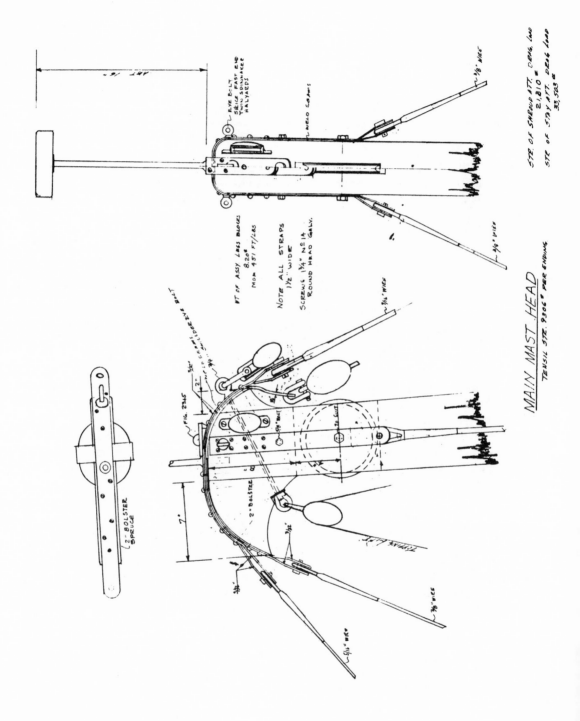

MAIN MAST HEAD

TENSIL STR. 9306 # PER ENDING

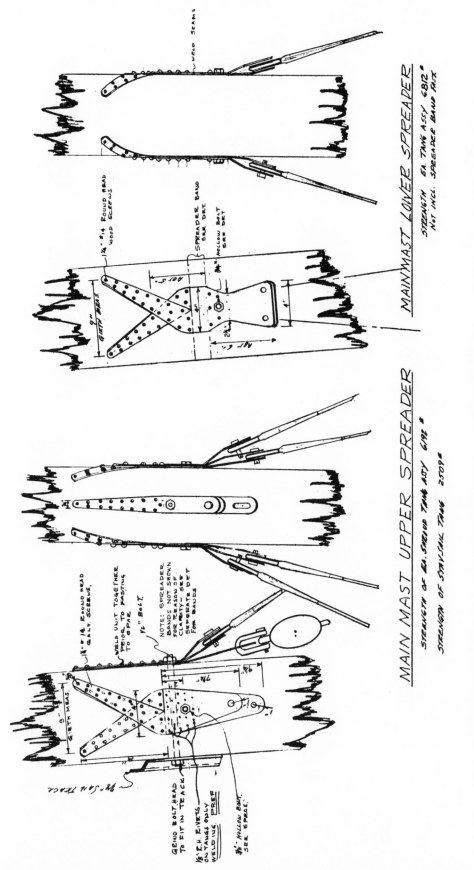

Figure 36. On these four pages: The rigging attachments used on the hollow wooden masts of two 45-foot ketch designs.

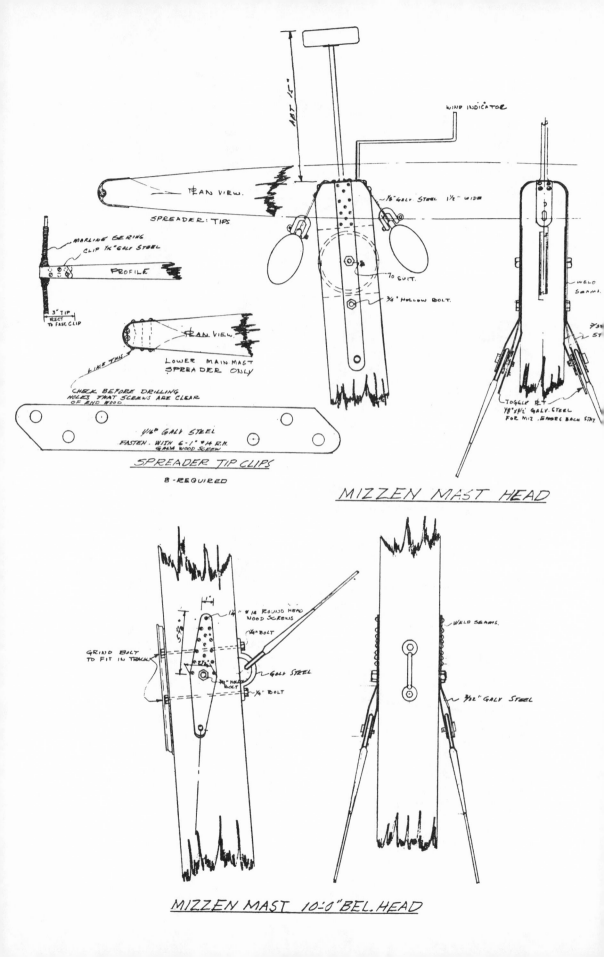

TRAN VIEW.

SPREADER TIPS

MARLINE SEIZING
CLIP 1/8" GALV STEEL

PROFILE

3" TIP
RECT
TO FAST. CLIP

LIKE THIS

TRAN VIEW.

LOWER MAIN MAST
SPREADER ONLY

CHECK BEFORE DRILLING
HOLES THAT SCREWS ARE CLEAR
OF END WOOD

1/16" GALV STEEL
FASTEN WITH 6-1" #14 R.H.
GALV WOOD SCREW

SPREADER TIP CLIPS
8 - REQUIRED

1/8" GALV STEEL 1 1/2" WIDE

TO SUIT.

3/4" HOLLOW BOLT.

WIND INDICATOR

WELD
SEAMS.

7/32"
ST

TOGGLE PL.
3/8"×1 1/2" GALV. STEEL
FOR MIZ. EMERG. BACK STAY.

MIZZEN MAST HEAD

14 ROUND HEAD
WOOD SCREWS

1/4" BOLT

GRIND BOLT
TO FIT IN TRACK

3/4" HOLLOW
BOLT

GALV STEEL

1/4" BOLT

WELD SEAMS.

3/32" GALV STEEL

MIZZEN MAST 10'-0" BEL. HEAD

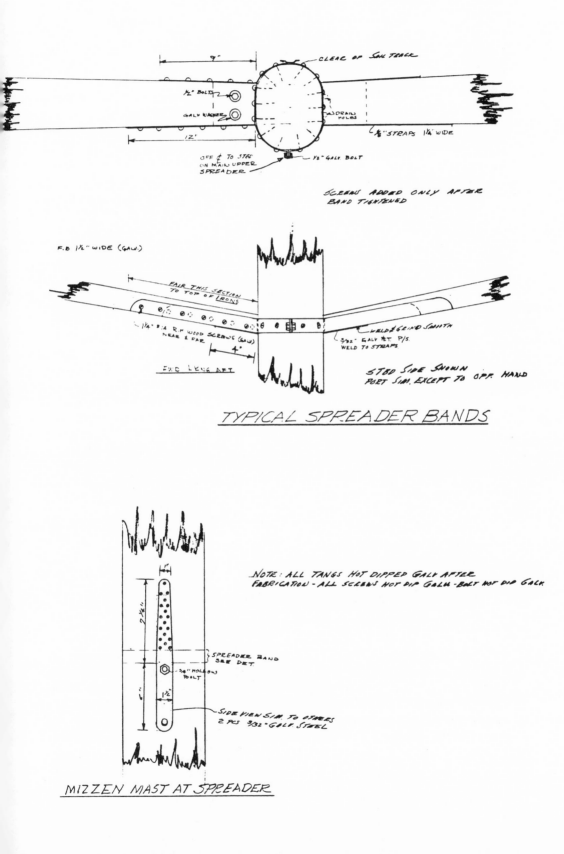

CLEAR OF SAIL TRACK

9"

½" BOLT

GALV WASHER

DRAIN HOLES

12'

OFF ℄ TO STBD
ON MAIN UPPER
SPREADER

½" GALV. BOLT

⅛" STRAPS 1¼" WIDE

SCREWS ADDED ONLY AFTER
BAND TIGHTENED

F.B 1½" WIDE (GALV.)

FAIR THIS SECTION
TO TOP OF IRONS

¼" Ø:4 R.H. WOOD SCREWS (GALV.)
NEAR & FAR

4"

FWD LENG AFT

WELD & GRIND SMOOTH

3/32" GALV MT P/S.
WELD TO STRAPS

STBD SIDE SHOWN
PORT SIM. EXCEPT TO OPP. HAND

TYPICAL SPREADER BANDS

NOTE: ALL TANGS HOT DIPPED GALV AFTER
FABRICATION - ALL SCREWS HOT DIP GALV - BOLT HOT DIP GALV.

1"

7½"

SPREADER BAND
SEE DET.

¾" HOLLOW
BOLT

6"

1½"

SIDE VIEW SIM. TO OTHERS
2 PCS 3/32" GALV. STEEL

MIZZEN MAST AT SPREADER

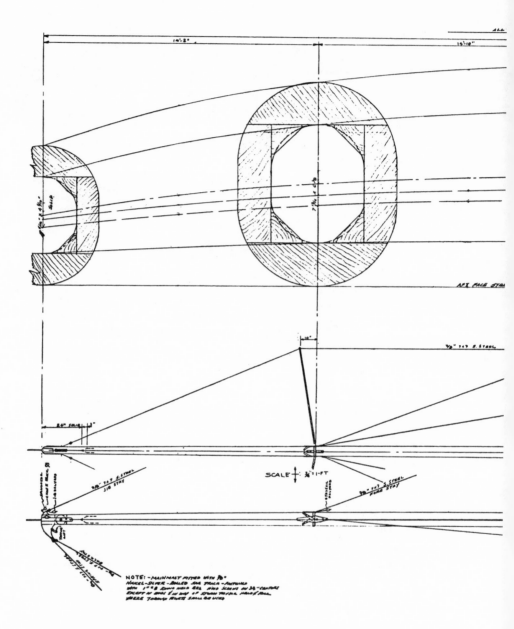

Building wooden spars was always a challenge and a joy for a skilled woodworker, and designing them for maximum strength and lightness required much study. Figure 37 is the plan for the mainmasts of the two 45-foot cruising ketches, which I built 30 years ago. After hard usage, including a circumnavigation, these masts still remain sound today.

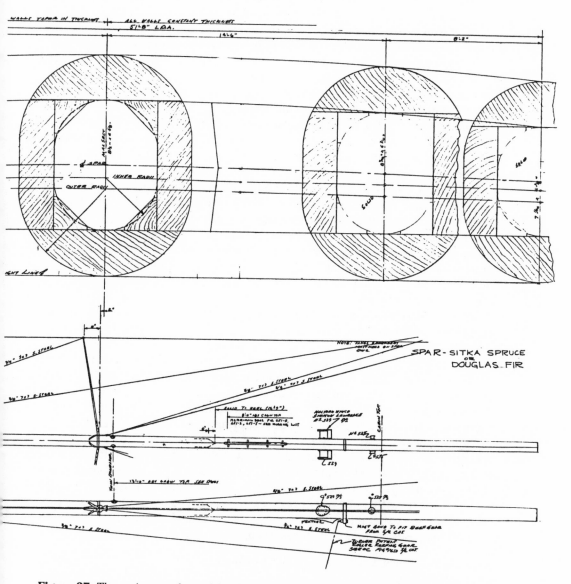

Figure 37. *The mainmast plan used with the two 45-foot ketches. After 30 years of use these wooden masts are still sound. The rigging attachments are detailed in Figure 36.*

Metal spars, when constructed in the builder's yard, are usually of standard mill sections in the form of tube or pipe. If the spars are large in diameter, the builder may have the material rolled, either with his own equipment or by jobbing it out. An alternative is to have the material spirally rolled; then, the finished section can be welded up by the builder or jobber. Without making the material-cutting operation too sophisticated, it is possible to use several tapers in the rolling process. Metals, especially aluminum, can also be spun to lesser diameters, as exemplified by aluminum light poles. For practical purposes, it would seem best to consider only those materials that are standard and can be worked by all builders.

The aluminum alloy pipes used for spars are 6061–T6 and 6063–T6, normally with a Schedule 40 wall thickness. In certain applications one can use Schedule 10, although the welding of this material is difficult with either Manual Metal Arc (coated electrodes) or MIG (Metal Inert Gas), unless the latter is of the pulse arc type. TIG (Gas Tungsten Arc) is the preferred welding method, but not always available in an otherwise well-equipped steel boatyard. In general, ³⁄₁₆-inch (.188-inch) and thicker aluminum presents no welding problems with proper joint preparation and cleanliness.

If, for reasons of strength, one should need to use Schedule 80 instead of Schedule 40 pipe to maintain a given diameter, it will often be found that steel tubing will provide the necessary strength and diameter at approximately the same weight as Schedule 40 aluminum pipe and at a fraction of the cost of Schedule 80. Steel tube is an excellent material for some spars in spite of the general notion that steel is heavier and therefore less suitable than other materials. With a bit of simple engineering (which it behooves the builder to do) one will find that he can very often realize a significant saving by using steel.

Mention should be made here that very often, especially in conversions, a new spar must be substituted for the existing one. Unless there was a stability problem to begin with, the builder should not substitute a lighter spar. In fact, on some vessels, an even heavier one may be in order, because a lighter spar will quicken the rolling period of the vessel. When the rolling period becomes too quick, the vessel becomes known as a "crank," and even the stoutest of stomachs finds her trying. (For the same reason, the old rule of thumb was to load two-thirds of the cargo in the lower hold and one-third in the 'tween decks. Following this rule usually resulted in a comfortable motion in a seaway.)

For comparative purposes, the standard sizes of aluminum pipe and steel tube are listed in Table 4. These are compared in all instances with the same diameter in a solid fir mast, using 35 pounds per cubic foot as the standard density of fir. It might seem from the table that one can always make an easy choice, but the true picture appears only with strength calculations. Masts restrained by shrouds are, for all practical purposes, columns under a compression load. The calculations for compression loading are based on Euler's formula, which is $P = \dfrac{\pi^2 EI}{L^2}$ for the maximum compression load of a pinned-end/pinned-end column, such as a mast stepped in a socket on deck with stays and shrouds. E is the modulus of elasticity; I is the moment of inertia of the section; L is the length of the supported column in inches; and P is the load in pounds. E for dense Douglas fir is 1,600,000 pounds per square inch (psi); for aluminum alloys, it is 10,000,000 psi; and for steel, it is 30,000,000 psi.

The moment of inertia for a solid spar is $I = \dfrac{\pi D^4}{64}$ and the moment of inertia for a hollow spar is $I = \dfrac{\pi(D^4 - d^4)}{64}$. D equals the outside diameter and d represents the inside diameter. Euler's formula applies to a pin/pin end condition.

TABLE 4.   Mast and Spar Materials

Solid Wood		Steel Tube			Aluminum Alloy Pipe		
Outer diameter (inches)	Weight (pounds/foot)	Outer diameter (inches)	Gauge	Weight (pounds/foot)	Nominal diameter (inches)	Outer diameter (inches)	Weight (pounds/foot)
4	3.05	4	11	4.97	3½	4	3.15
4½	3.86	4½	11	5.61	4	4½	3.73
4¾	4.31	4¾	11	5.93			
5	4.77	5	11	6.25			
5¼	5.26	5¼	11	6.58			
5½	5.77	5½	11	6.90			
5⅝	6.04	6	11	7.54	5	5⅝	5.05
6	6.87	6⅝	11	8.68	6	6⅝	6.56
6⅝	8.38	6⅝	10	9.32			
		8	11	10.77			
8	12.23	8	10	11.58			
		8	7	15.03			
8⅝	14.20	10	10	14.50	8	8⅝	9.87
10	19.09	10	¼"	26.03			
		12¾	3⁄16"	24.17			
10¾	22.76	14	10	19.84	10	10¾	14.00
12¾	31.03	16	10	22.71	12	12¾	17.14
14	37.42						
16	48.87						

Now if the mast passes through partners and steps on the keel, the heel is considered fixed and the other end pinned, and this fixed/pin condition has twice the strength of a pin/pin condition. The third possibility is a free-standing spar that steps through the deck and is without shrouds or stays; this, in reality, is a cantilevered beam. It has one-fourth the strength of a pin/pin configuration or one-eighth the strength of a fixed/pin column, and is subject to bending rather than compression loads.

From Table 4, using an outside diameter of 4½ inches as an example, a solid wood spar has an I of 20.13, aluminum alloy pipe has an I of 7.56, and steel tube has an I of 4.11. The maximum compressive load on a 20-foot spar would be as follows (the strengths and weights given are relative to wood):

Material	Pin/Pin (psi)	Fixed/Pin (psi)	Fixed/Free (psi)	Bending Strength	Weight
Wood	5,519	11,038	1,380	1.0	1.0
Aluminum pipe	12,954	25,908	3,239	0.97	2.35
Steel tube	21,127	42,254	5,282	1.45	3.83
Steel tube*	43,117	86,234	10,779	1.45	3.83

*According to S. Timoshenko (Theory of Elastic Stability, McGraw-Hill Book Co., New York, 1936), one should use $(0.70L)^2$ when calculating the compressive strength of steel cylinders, which yields an even greater strength for steel.

Once a column has been loaded to its maximum, it will begin to bend, and in the case of metal, it will next buckle, so the prudent builder should also investigate these two loadings. Instead of buckling, wood will break.

Since builder-fabricated metal spars are usually round, they may also be tapered in much the same manner as round wooden spars. The metal spars normally have a smaller diameter than the same spars made of wood, and they therefore need less extensive tapering. Tapering is done by cutting wedges from the spar, their length being the same as that of the tapered section. For a uniform (straight) taper, the untapered circumference of the spar is divided into sixths or eighths, and longitudinal segment lines are laid down on the spar. The circumference of the spar at the tapered diameter is then divided by the number of segments; half of the resultant value is laid off on either side of each segment line at what is to be the tapered end. The cuts are made as in Figure 38 (A), where all the wedges removed have straight sides, and the segments are then squeezed together and tack welded where they touch. A compound taper, using the standard spar diagram as shown in the lefthand sketch in Figure 30, is shown in Figure 38 (B), where all wedges removed have a curved edge. A finished taper on all sides except one is also shown (C), where all wedges removed have a curved edge identical to the compound taper; in this taper, however, the after segment is held as a straight line continuation of the uncut spar, and all other segments are pulled down to it. This taper requires that heat be applied so that the root of each segment can be edge-set. In the top view of each of

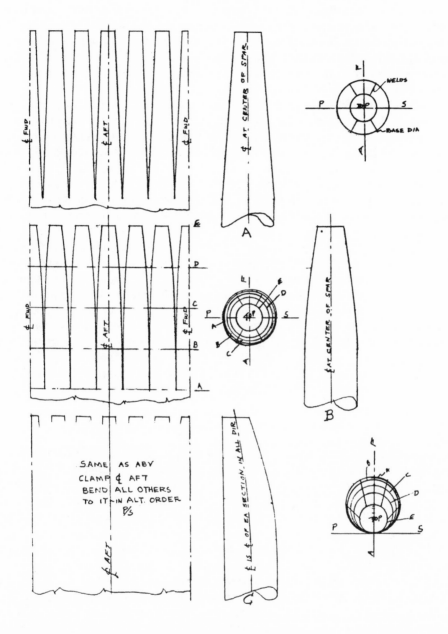

Figure 38. *A metal spar is tapered by cutting wedges having the same length as the tapered section. Removing straight-sided wedges results in a uniform taper (A). Wedges with curved sides give a compound taper (B) or, if the after segment is held as a straight line in profile, the taper shown in* C.

these spar tapers (Figure 38), the letters *A, B, C, D,* and *E* refer back to the mathematical diagram for laying out the taper, shown on Figure 30. The apex of the cut must end in a hole about ³⁄₁₆ inch in diameter; otherwise, the weld will tend to crack and propagate into the uncut material, especially with the aluminum alloys. The illustrated cuts are self-explanatory, showing the spar flattened out and in section for clarity.

The cut-and-weld method does not produce a true round; however, it does produce a close approximation. Aluminum spars using Schedule 40 pipe may be sanded round without a significant decrease in strength. In steel spars, only the welds should be ground smooth and slightly rounded. Excessively tapered lengths are not feasible using a round section, due to loss in strength of the as-welded section in aluminum alloys and (since the welds can be made from only one side) the inability to achieve more than an 85 percent weld efficiency in steel. If total taper is required, one can accomplish it in an oval or semirectangular spar section provided a brake press is available.

In aluminum alloy spars, the joints should be made with an internal sleeve, its length about three times the diameter of the pipe or tube. The sleeve may be attached to the individual sections with either machine screws or rosette welds. Tapered aluminum alloy light poles used as extensions to the lower portion of the mast offer the least expensive way to achieve a reduction in mast diameter and require only that the joint between the pole and the lower portion be at a slight angle *if* the after side of the mast must be straight to accept sail track. The smaller light poles make fine topmasts in vessels that require them and can be used as manufactured, requiring only the addition of a cap and thumbs for the rigging.

CHAIN

Most people associate chain exclusively with anchoring, but it has many other uses on sailing and motor vessels, such as for chain slings, cargo lashings, nets, trawling door bridles, net bridles, and heavy lifting. On sailing vessels, chain is used for the bobstays, bowsprit shrouds, and jib bridles. Chain is also used sometimes to raise the turnbuckles well off the deck to prevent damage from deck cargoes and keep them from being immersed on deeply laden vessels. Aloft, chain has several uses, principally in sling chains for the yard trusses and heel chains connecting the mast cap and the topmast heel.

Table 5 indicates the sizes of chain commonly used and some of its characteristics. The figures are for galvanized chain. The primary information needed to select chain is the diameter, weight, and working load. Manufacturers use many different nomenclatures in selling chain, and link size will vary from one manufacturer to another; however, the material diameter remains basically the same. Chain should never be loaded beyond its working strength.

The minimum proof test is twice the working load for proof chain, and 1.64 times the working load for high-test chain. The minimum breaking strength is four times the working load for proof and three times the working load for high-test. It will be noted, then, that the minimum breaking strength for high-test chain is 1½ times

TABLE 5. **Characteristics of the Commonly Used Sizes of Galvanized Chain**

| Chain size (inches) | Proof chain, grade 30 | | High-test chain, grade 40 | |
	Weight (pounds/foot)	Working load (pounds)	Weight (pounds/foot)	Working load (pounds)
³⁄₁₆	0.40	750	—	—
¼	0.71	1250	0.75	2600
⁵⁄₁₆	1.07	1900	1.11	3900
⅜	1.58	2650	1.57	5400
⁷⁄₁₆	—	—	2.13	7200
½	2.68	4500	2.64	9200
⅝	4.10	6900	4.09	14000
¾	5.80	9750	6.03	19750
⅞	8.11	11375	—	—
1	10.40	13950	—	—

Source: American Chain & Cable Company, Inc.

greater than that of proof chain. The two grades differ in link length: in ⁵⁄₁₆-inch grade 30, the links measure 1.10 inches, while in grade 40 they measure 1.01 inches. Grade BBB chain, sold by some companies, has a slightly different link length and working load. It is, however, a calibrated link length and is often specified as a standard by manufacturers of windlasses in the United States. Therefore, when it is important that the anchor chain fit a wildcat, the builder should send the windlass manufacturer a sample of the chain. The use of high-test chain for anchoring smaller vessels gives a weight saving for the equivalent strength, and the smaller diameter stows in less space. The usual practice is to use one size less for high-test than would have been used with proof, since the difference in yield strengths does not warrant a greater reduction in most instances.

Shackles should always be of the same material (alloy) as the chain on which they are to be used. When shackles are used for attaching the chain to the anchor, the usual practice is to use a swivel shackle on the first link, which then permits the use of a much larger shackle to the ring of the anchor.

TURNBUCKLES AND DEADEYES

Turnbuckles are the most common fittings used to tighten the standing rigging. Commercial vessels use galvanized iron (steel) turnbuckles, while yachts use bronze or stainless steel. Turnbuckles come in two varieties: the open type, where the threaded portion shows inside the body of the coupling, and the barrel type, which conceals all threads within the coupling barrel. The latter type is to be preferred with galvanized turnbuckles, since a grease fitting can be fitted near the center of the barrel, which lessens the possibility of the threads seizing. With the open pattern, it is customary on commercial craft to fill the body with waterproof grease, wrap it with canvas, and then cover it with a sewn canvas sleeve. In order to prevent the body from turning, the galvanized open turnbuckles are fitted with check nuts; the barrel type has a hole drilled through the center of the body that permits a wire to pass through. A sheer pole is added at the upper eye or jaw to keep the rigging from exerting a twist on the threads. Many of the yacht-type turnbuckles have holes drilled near the ends of the screws, into which a wire or cotter pin can be rove to prevent the body from turning. If the turnbuckle lacks this feature, then check nuts are used. In a steel vessel it is easy to align the chainplate eyes fair with the rigging leads, working directly from the rigging plan, which eliminates the need for toggles or shackles to compensate the turnbuckle for misalignment. Turnbuckles used at sea have various terminal endings: jaw/jaw, jaw/eye, and eye/eye. The open hook type is not used, as it has less strength and is subject to fouling. Table 6 indicates the strengths and working loads for galvanized turnbuckles.

TABLE 6. Strengths and Working Loads for Galvanized Turnbuckles

Diameter over threads (inches)	Breaking strength (pounds)	Working strength (pounds)
¼	1350	250
⁵⁄₁₆	2250	500
⅜	3350	660
½	6250	2000
⅝	10000	3000
¾	15000	5000
⅞	21000	7000
1	27500	9000
1⅛	34800	11500
1¼	42900	14000
1½	61800	20400
1¾	84200	27800
2	110000	36300

The weight of the turnbuckles will vary according to the end fittings. As a guide, a 1-inch turnbuckle weighs about 13 pounds and a 2-inch turnbuckle weighs about 100 pounds. There is no point in cross-referencing the turnbuckle with wire diameter, since the strength of the wire depends on its material and construction. Wire can only be compared with other wire. The working strength of the wire chosen then determines the turnbuckle size. The table reflects the use of drop forged steel, with a safety factor applied first to the steel and then to the turnbuckle, so that the approximate total safety factor is 5 to 1. The turnbuckles should all be installed in such a way that, in order to tighten them, one always turns the body in the same direction. Overtightened turnbuckles impose a tremendous compression load, which then weakens rather than strengthens a mast or spar.

Deadeyes are the alternative to turnbuckles, and, except in the highly stressed yacht rigs, they are superior because they do not rust or corrode. They also yield just a bit when a vessel rolls in a seaway, thus eliminating most of the jerks and shock loads associated with turnbuckles; they require less maintenance; they are self-aligning; and they cost less. In the days when tarred hemp was used, the lanyards would stretch for the first few months after being set up, and would have to be set up again, which could not be done correctly at sea. When the lanyards were properly maintained by tarring, however, they caused little trouble after the first year. Nowadays, people use synthetic polyester rope in lieu of hemp, and the lanyards seldom need further adjustment after the initial sailing trials. Deadeyes are often connected to each other with a turnbuckle, with the lanyards rove off to camouflage and hide it. This is fakery and an admission of inept seamanship, neither of which has any place on a proper seagoing vessel. In the larger vessels, marline-clad 5 x 19 rope of improved plow steel, fiber core, can be used to advantage for the lanyards, since it has several times the strength of polyester for the same diameter. The separation between the upper and lower deadeyes should measure not less than 18 inches for the smallest vessel, with 30 inches being about the maximum for larger vessels.

WIRE ROPE

With the exception of a few developing countries, wire rope has become the universal standard for standing rigging. Wire rope is made of different sizes of wire, and the individual wires may be stranded around a center in either single or multiple operations; the wires in a strand number from seven to more than 50. The strands converge in a die at constant pressure to obtain a constant diameter. After that, the strands are taken up on a spool, then helically preformed before passing to the "closer," which lays them helically around the core, forming a finished wire rope.

With respect to wire rope, as with respect to so many other things, commercial and yacht practices differ. Commercial vessels use galvanized wire rope, while yachts normally use stainless steel. In yachting, one need not do much, if anything, to the standing rigging. In fact, there is little one *can* do, since any lubrication is apt to stain the sails. Unfortunately, while stainless steel is corrosion-resistant, it does corrode and it does stain. Furthermore, the stainless steel wires will not tolerate being starved of oxygen, and this is but another inherent form of corrosion associated with this material. Stress corrosion results from the cold working of the metal and is the most frequent cause of failures. Brown discoloration gives the telltale warning. Since the terminal ends are usually swaged (cold working) due to a preference for using 1 x 19 wire construction (1 wire, 19 strands) to reduce diameter and thus windage, the corrosion problem is inherent from the beginning and not something that just develops "someday." Yacht rigging is laid up dry, so there is friction along the wires as tension increases or decreases (cold working), and the absence of lubrication leaves nothing to ward off contaminants—salt water and dirt. Failure does not take place gradually, but immediately and without warning.

Galvanized wire rope, on the other hand, is lubricated during the manufacturing process. The galvanized wire rope gives a long warning before failure occurs, such as rust seeping through the strands and the breaking of individual wires, forming so-called fish hooks. To preserve galvanized wire rope one simply coats the wire once a year or at the end of each voyage, depending on the conditions. If wormed, parceled, and served, galvanized wire will have an almost indefinite life. Indeed, I have reused wire, so-treated, that had already seen 40 years of service, and with complete satisfaction. One can lubricate the wire with fish oil, tung oil, boiled linseed oil, Stockholm tar, Oriole Pine Tar (found in the U.S.), white lead mixed with tallow, or aluminum pigment mixed in an oil base. On the smaller vessels, parceling is done with cloth (tarred) electrician's tape, since the wires are small. For larger wires duck soaked in pine tar is used. In any case, one should never use plastic tape, for it will trap any moisture that may seep into the strands. The use of marline of Russian or Italian hemp has been more or less discontinued in favor of three-strand tarred nylon of the same diameter.

TABLE 7.   Strength (short tons) of Wire Rope as a Function of Size and Construction

Diameter (inches)	1 x 19	7 x 7	6 x 7	8 x 19	6 x 25
$\frac{1}{32}$	0.075*				
$\frac{3}{64}$	0.14	0.134			
$\frac{1}{16}$	0.25	0.25			
$\frac{5}{64}$	0.40	0.33			
$\frac{3}{32}$	0.60	0.46			
$\frac{7}{64}$	0.80	0.63			
$\frac{1}{8}$	1.05	.85			
$\frac{5}{32}$	1.65	1.30			
$\frac{3}{16}$	2.35	1.85	1.61		
$\frac{7}{32}$	3.15	2.40			
$\frac{1}{4}$	4.10	3.05	2.84	1.80	1.80
$\frac{5}{16}$	6.25	4.60	4.41	2.80	2.80
$\frac{3}{8}$	8.40*	6.55	6.30	4.10	4.10
$\frac{7}{16}$	—	—	8.52	5.50	5.50
$\frac{1}{2}$	—	—	11.10	7.25	7.25
$\frac{9}{16}$	—	—	14.00	9.25	9.25
$\frac{5}{8}$	—	—	17.10	11.50	11.50
$\frac{3}{4}$	—	—	24.40	16.00	16.00

Available in stainless steel only.

Source: Universal Wire Products, Inc., North Haven, Connecticut

The cores of wire rope may be 1 x 7 strand, 1 x 19 strand, independent wire rope core (IWRC), and fiber core. The grades of wire rope are improved plow steel (IPS), extra improved plow steel (EIPS), double extra improved plow steel (EEIPS), and galvanized aircraft (GA/C). There are other grades of wire as well, such as stainless steel, designated by alloy, and also bronze rope. I mention this because a designer will often use the abbreviated symbols on his drawings rather than write it all out. The lowest grade used for standing rigging is improved plow steel with an independent wire rope core.

Table 7 indicates the diameter and strength (short tons) for the various constructions of wire rope used for standing rigging. The table includes 8 x 19 and 6 x 25 wire construction because these are used as elevator ropes and are often available free or at scrap metal value after they have been condemned and replaced by new cable. The elevator companies replace them long before the wire shows any sign of damage. Sometimes a company has a supply of what it considers "shorts," which may be long enough for a shroud or a stay. I know very well that these cables are used for standing rigging, but I will always prefer galvanized 7 x 7 or 6 x 7 wire rope. The smaller sizes of wire listed are used as control wires and aerials.

Running rigging of wire rope requires flexibility and minimum stretch. For steering cables and halyards on yachts, the 7 x 19 is the most popular, in sizes up to ⅜-inch diameter, and is usually of stainless steel. Larger vessels use 6 x 37 fiber core for cargo whips. Fishing vessels—trawlers in particular—probably make the most extensive use of wire.

Wire rope must pass over sheaves of much larger diameter than required for fiber rope, or the wires will be deformed. The same holds true of the drums upon which the wire is wound. The minimum diameter of sheaves is 14 diameters of the wire for tiller tope (6 x 42), and 42 diameters for trawl rope (6 x 7). Lesser diameters are often used, but one must then accept a correspondingly shorter rope life.

TERMINAL FITTINGS

To be at all useful, wire rope must have its terminal ends attached to or encompassing some sort of fitting. This fitting may be an eye to which another fitting is hooked or pinned, or a clevis that attaches to the eye of another fitting.

I have already spoken of swaged terminal ends used in yacht rigging with 1 x 19 wire rope. With this method, either a clevis or an eye terminates the length of rigging, and is put on by a machine that squeezes and at the same time elongates a stainless steel sleeve over a short length of the wire end. The swaging results in a streamlined fitting that blends into the wire with a minimum transitional difference to the wire diameter, and it is one of the lightest terminal fittings available. On the other hand, the transition point is a weakness unless the fitting aligns perfectly with the wire; furthermore, stainless steel when worked (squeezed) initiates some stress corrosion, and these fittings are not reusable.

Another terminal end fitting is the poured socket, which may be either a clevis or an eye. The wire rope is inserted into a hole of the correct diameter, the strands are unlaid, and the individual wires are spread, deformed, or both. Lubricants are cleaned from the wires, and the wires are tinned. Then, the casing of the fitting is preheated and the socket poured full of molten zinc. Some riggers use babbitt because of its lower melting point; however, it is inferior to zinc. Such terminal endings have as much strength as the wire, require a minimum of equipment, work with any of the wire constructions, can be reused, and have an excellent finished appearance. The primary disadvantage is the weight of these endings when compared to others, especially if the long sockets are used.

Many brands of terminal endings do not require swaging but instead use studs, cones, and the like, inserted into the core of the wire. These are sold under proprietary brand names and trademarks such as Sta-Lok, Norseman, Seatest, and Castlok, and will work with 1 x 19 and many of the other wire constructions, except those with a fiber core. The end result has the appearance of a swaged fitting, but they can be made up with the hand tools normally found aboard a vessel. The fittings are reusable and do not have the problems of stress corrosion associated with swaging.

The most common ending of all is made by bending the wire around a thimble and securing the wire ends to the standing part of the wire, using wire rope clips, splices, seizing, or splicing sleeves. Motor vessels often use the wire rope clips, and several well-known cruising sailors recommend them as the best way to secure wire rigging. Most vessels, however, carry them only for use in emergencies. When properly installed, there is a minimum of damage to the wire. The clips are reusable, and can be installed using just a wrench; however, they are prone to fouling and can tear the sails.

Splicing is the weakest method of terminating the wires and requires a skill common among able-bodied seamen, who are hard to find today. The splice, when served, is a handsome ending and diminishes into the standing part of the wire. The simplest method of forming an eyesplice around a thimble is by using a three-jawed rigger's vise; otherwise, the wire must be temporarily seized to the thimble. Some riggers prefer to make the splice and then pound in the thimble, which can cause damage to both the wire and the thimble. Others insert the thimble after the first tuck. While all stranded wire can be spliced, the most difficult is 1 x 19.

An exception to the diminished strength of a spliced wire is the French splice, which is the same strength as the wire when finished; however, it cannot be worked around either a thimble or a deadeye, and is reserved for larger eyes, such as those that pass over the masthead. This splice does not harm or deform the wire, so may be undone and the wire reused.

In the larger sailing vessels, it is a common practice to lay up the end of the wire for a distance of 18 to 36 inches after it has passed around the thimble, then seize it to the standing part. This is a strong method that does not weaken the wire, and the wire is reusable.

Nicopress splicing sleeves, constructed of zinc-plated copper or aluminum, are shaped like a figure eight when viewed end-on. The standing part and the end of the wire are inserted into the sleeve, and the sleeve is then crimped together using special hand tools in the smaller sizes or a hydraulic press and special dies in the larger sizes. These tools extrude the sleeve by reducing the diameter. The Nicopress fitting can be used with any wire configuration and is quick, and the strength of the sleeve exceeds that of the wire. The sleeves and the wire in way of the sleeves are not reusable, however. The British equivalent of the Nicopress sleeve is the Talurit swaging process.

There are two types of thimbles—the ordinary open type and the solid type. The former is the most common and easiest to obtain, but for use at sea for standing rigging it is inferior to and makes a poor substitute for the solid thimbles. With the open thimbles the loading is concentrated on one small area, which can be deformed long before the wire reaches its minimum breaking strength. When this happens, the wire kinks, which further reduces its strength. The pin of a shackle, clevis, or turnbuckle, when inserted through the eye of the thimble, rests on a rounded surface, and when two rounded surfaces come in contact with each other, little of the total material comes to bear. This stresses the pin in a way that was not intended. Then, too, if the eye is large, a shackle can upset within it, placing a strain on the side rather than the center of the thimble. The solid thimbles, on the other hand, have a hole for the shackle pin near the neutral axis of the material, and cannot be deformed. A pin inserted through the hole has a bearing surface width in excess of the wire diameter, so the possibility of the thimble fouling is nil. The source for the solid thimbles is Harris-Walton Ltd., Two Woods Lane, Brierly Hill, West Midlands, DY5 1TR, England. In a properly rigged vessel, the ordinary thimble should be restricted to rope.

RATLINES

Ratlines are the traditional means of going aloft and provide the only way on a gaff-rigged vessel without resorting to a bosun's chair or climbing the hoops. Vessels rigged with sail track often have steps on the forward side of each mast or on the sides of the masts. Ratlines should be spaced from 13 to 16 inches apart, depending on the steepness of the angle: the steeper the angle, the greater the space. There are several ways to make ratlines; the type used will depend on the size of the vessel and how often the crew must use them. All-rope ratlines set up in the usual fashion are possible only on the larger vessels, the shrouds of which are large enough in diameter so that the tension required to set them up exceeds the tension required to squeeze them together under the weight of a man. Rope ratlines are eye-spliced at each end, seized to the forward shroud (the swifter), secured to each succeeding shroud with a clove hitch (on the outside of the shroud), and seized to the last shroud. Vessels with three lower shrouds per side customarily have ratlines on all three; however, with four shrouds, only every fifth ratline is carried through to the after shroud. On schooners, this is called a "catch ratline." Square-riggers have the catch ratline carried to the swifter. The eyesplices are always seized in the horizontal position. When crew go aloft primarily for overhauling the gear, the ratlines are often set up on only one side of each mast—for instance, on the starboard side of the foremast and the port side of the mainmast.

Figure 39 shows a method of using only rope for ratlines on the smaller vessels. A large quantity of rope is required; however, the weight, windage, and chafe are minimal. This method seems most suitable when only one side of each mast is to be rattled down.

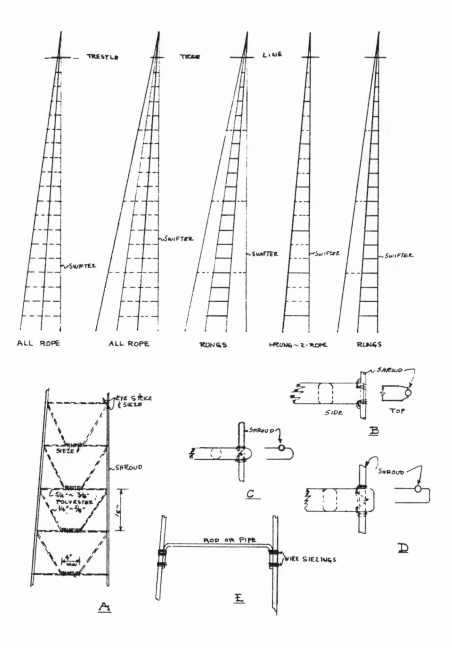

Figure 39. *Ratline configurations and details.*

Most small vessels with only two shrouds must resort to the use of wooden rungs in lieu of rope, to keep the rigging from squeezing together. On larger vessels, where crew frequently need to go aloft to handle the topsails when tacking or to furl the squaresails, wooden rungs are also used. With three shrouds, they are fitted between the swifter and center shroud, with a rope catch ratline attached to the after shroud. This eliminates the problem of sails chafing on the rungs. With four shrouds, the rungs attach to the two middle shrouds, with catch ratlines run to both the swifter and after shrouds. When the rigging wire equals or exceeds ⅜ inch, a rung every third ratline will suffice and saves weight. Figure 39 also shows these methods. Oak, ash, and other hardwoods make the best rungs. The shape depends on the span: for short spans, round is satisfactory, while the larger spans will require rectangular sections. Whichever type is selected, it is used for all the rungs on the vessel. The ends of the rungs can be scored for the wire when using rectangular sections (B); the side near the ends can be scored when using either rounds or rectangular sections (C and D). The latter is the most common method. Frequently one sees U-bolts encircling the wire with the nuts set up on the outside of the rung; this method can harm the wire and is more a landsman's way of solving a seaman's problem. Round iron bar in the form of a staple is also used (E), seized to the shrouds with wire. This style puts considerable weight aloft, so aluminum rods are sometimes substituted, but these require a greater diameter than iron. When ratlines are used, it is customary to serve the shrouds, which increases the diameter, making them a better and kinder handhold, but also increases the windage.

CORDAGE

Sailing vessels of less than 100 feet usually use only natural or synthetic rope for sheets, halyards, purchases, and other running rigging, the exception being yachts with jibheaded sails, which use wire halyards wound on winch drums attached to the mast. Rope size is specified in the U.S. by diameter; elsewhere, it is specified by circumference. To convert from one to the other, one basically either multiplies diameter or divides circumference by three. The plans usually specify certain rope sizes, but if they do not, ensuring good handling characteristics is more important than keeping the diameter to a minimum. For example, consider the selection of the mainsheet for a small schooner with a 600-square-foot mainsail. A twofold purchase is used, giving a theoretical advantage of 5:1; taking into account friction, the actual advantage is 4.1:1. Assuming the maximum gust that could be encountered is a "very strong breeze" (4 pounds per square foot), then the total load on the block at the boom is 2,400 pounds, which, divided by 4.1, equals 585 pounds on the fall. The working load of Manila rope is one-fifth and of polyesters one-ninth the tensile strength; therefore, one needs 2,925 pounds of strength for Manila (⁹⁄₁₆ inch in diameter) or 5,265 pounds of strength for polyester (⁷⁄₁₆ inch in diameter). Rope of ½-inch diameter is the minimum that one can grab and heave on without cutting his hands, and for a sail of the size mentioned, ⅝-inch is better. The ⅝-inch rope has more than adequate strength for the loads imposed and is kinder to the hands, and so would be the better choice. I hasten to add that normal design practice is to select the sheet that is subject to the worst loading and then to average its loading with the loadings of the other sheets to arrive at an acceptable mean, since the economics lie in using the same size rope for all sheets and halyards.

The use of anything other than synthetic rope in yachts has become almost unknown. Perhaps this is because synthetic rope can be neglected; because it retains its color; and because cost is secondary. Perhaps it is just a fad. There is no question of its value in the larger sizes, where both strength and diameter count. Nor can one doubt the shock-absorbing powers of nylon anchor rodes. However, all synthetics become brittle with exposure to sunlight, and they all age by being continually exposed to the elements. In commercial vessels, synthetic rope has some uses, and it is employed whenever it will do a superior job; otherwise, Manila rope is used, not only for economic reasons (it costs about one-half as much as polyester), but because it is easier to work with, holds splices and knots better, permits easy replacement of worn strands, and has infinite end uses, including bag-o-wrinkles, fenders, chafing gear, and swabs.

Braided polyester and combination polyester/nylon rope are commonly used for sheets and halyards on yachts because the line is perfectly round, easy on the hands, works well with most yacht fittings, stretches very little, has high abrasion resistance, and will not kink or hockle. The splicing is not as straightforward as with three- and four-strand rope, but it is not difficult to learn. One wonders why it is not, therefore, universal? While braided line can be spliced, the types of splices are very limited, the types of knots that can be used with it are even more limited, and it is difficult to repair. Repairs of any magnitude require replacement of the damaged section, while with regular rope, just the affected strand requires replacement. To make splices aloft is difficult in the extreme because of the splicing tools needed. On the other hand, multibraid nylon or polypropylene, stranded and woven, is used for anchor and mooring lines, especially on the larger vessels, since these lines have excellent shock-absorbing qualities and they can be stowed wet, which would be unthinkable with Manila.

Table 8 gives the sizes and tensile strengths for the various types of rope.

BLOCKS

Wooden blocks are designated by shell length and are usually made of ash, elm, or lignum vitae, although other woods, such as teak, oak, silver bali, delmari, and dense mahogany are used occasionally. Wooden blocks may or may not have a becket located below the breech. They come with numerous top fittings, including front- or side-facing hook, lashing eye, spans, Coleman hook, sister hook, and shackle. The shackle may be normal, twisted, or upset. The sheave bearings are either iron-bushed, patent roller–bushed, or Oil-Lite–bushed. Blocks for wire rope are designated by the sheave diameter at the bottom of the groove (B.O.G.), not the diameter over the rim, and they have metal shells. Table 9 indicates the sizes of wooden-shell blocks used on vessels up to 80 feet on deck.

TABLE 8. Tensile Strengths of Natural and Synthetic Ropes

Size (inches)		Manila	Polypropylene*	Nylon	Dacron** (polyester)
Diameter	Circumference	Tensile strength (pounds)	Tensile strength (pounds)	Tensile strength (pounds)	Tensile strength (pounds)
³⁄₁₆	⅝	405	800	1000	1000
¼	¾	540	1250	1650	1650
⁵⁄₁₆	1	900	1900	2550	2550
⅜	1⅛	1215	2700	3700	3700
⁷⁄₁₆	1¼	1575	3500	5000	5000
½	1½	2385	4200	6400	6400
⁹⁄₁₆	1¾	3105	5100	8000	8000
⅝	2	3960	6200	10400	10000
¾	2¼	4860	8500	14200	12500
¹³⁄₁₆	2½	5850	9900	17000	15500
⅞	2¾	6930	11500	20000	18000
1	3	8100	14000	25000	22000
1¹⁄₁₆	3¼	9450	16000	28800	25500
1⅛	3½	10800	18300	33000	29500
1¼	3¾	12150	21000	37500	33200
1⁵⁄₁₆	4	13500	23500	43000	37500
1½	4½	16650	29700	53000	46800
1⅝	5	20250	36000	65000	57000
1¾	5½	23850	43000	78000	67800
2	6	27900	52000	92000	80000
	working load	⅕	⅙	⅑	⅑
	working load %	20	17	11	11

*Polypropylene is monofilament.

**Du Pont registered trademark.

Source: Wall Rope Works–New Bedford Cordage Co.

TABLE 9. Sizes of Wooden-Shell Blocks Used on Vessels Up to 80 Feet on Deck

Rope Diameter (inches)	Shell Length (inches)
⅜	3
½	4
⁹⁄₁₆	5
⅝ or ¾	6
¾	7
⅞	8 or 9
1	10 or 11
1⅛	12 or 13
1¼	14 or 15

The weight and maintenance of wooden-shell blocks has all but eliminated their use in yachts, with the exception of the "character" types that need them for appearance' sake. The usual block is now made of sheet stainless steel, aluminum, Bakelite, nylon, or linen laminates. The blocks are not only lighter, but need little in the way of maintenance. Although their first cost is higher, their useful life and better bearings, which can tolerate immersion in salt water without freezing, make them less expensive in the long run, so they can be found on many commercial vessels. Wooden blocks should be overhauled at least once a year, the pins examined and replaced if worn, the bushings oiled or greased with anhydrous lanolin, and the strops inspected, painted, and bedded.

Enough goes wrong on every new vessel and continues to do so for several months after delivery. The problems are mostly minor in nature, but vexing nonetheless. A seaman can accept problems of this sort, but slovenly workmanship in rigging cannot be justified.

4

▽

LIFE RAFTS,
BOATS,
AND DINGHIES

No vessel should venture on long ocean passages without some sort of lifesaving devices. The vessel's boat (dinghy) provides only marginal safety, since this type of craft is made more for working than for saving lives. Furthermore, it is usually not feasible to keep survival stores in the dinghy at all times. Indeed, on a passage, most vessels have their boats filled with miscellaneous gear that would have to be cleared away before any survival gear could be loaded. On some of the larger vessels with room for more than one boat, it is not uncommon to have a proper dory fitted out as a lifeboat prior to departure, but the dinghies and many inflatables normally associated with yachts are usually inadequate in size and design. Inflatables deflated and stowed below decks or bagged on deck for the passage cannot be considered as lifesaving devices, because inflating them with a foot pump is laborious and time consuming, and may have to be done when circumstances make time of the essence.

LIFE RAFTS AND LIFEBOATS

Larger vessels customarily have lifeboats and the proper davits to handle them, conforming to SOLAS (International Convention for the Safety of Life at Sea) regulations, and of a type approved by the United States Coast Guard. The smallest lifeboat manufactured is a 12-foot model with a capacity for six persons; however, the minimum length with special approval for international voyaging is 16 feet.

Smaller vessels seldom enjoy the luxury of a lifeboat and must make do with one or more inflatable life rafts. The rigid rafts are approved by SOLAS but not the USCG, which is fortunate, for they are heavy and awkward. Life rafts are manufactured in sizes that will accommodate 4, 6, 8, 10, 12, 15, 20, and 25 persons, but the four-person size is not approved for ocean service. The capacity of a life raft must equal or exceed the number of persons aboard. Larger capacity than the minimum gives everyone a bit more room, which all hands will welcome after a few days of floating at sea.

The rafts are stowed in fiberglass containers, and the builder will be required to construct suitable cradles for them. Rafts must be located where they will be easy to release, will come clear, will not foul in the rigging, and will float free in the event the crew has insufficient time to release them. Most sailing vessels have trouble meeting these requirements. Unquestionably, rafts occupy a lot of valuable space, but careful study of the options often suggests the least objectionable site. The ideal, of course, is to have two rafts, but small commercial vessels or yachts usually cannot justify the expense. With only one raft, the ability to launch from either side of the vessel is a prerequisite, as hopefully there will be a lee. The notion that a raft can simply be cast over the side and stepped aboard with added rations and water sounds fine to armchair sailors and textbook strategists who have never had to abandon ship except in drills during calm weather or at dockside. "Dey wasn't dere, Charlie."

Life floats constructed of canvas-covered balsa are occasionally seen on the smaller commercial vessels as a secondary safety provision in the event that the life raft fails to function. These floats have a platform suspended via netting below them. In tropical waters this is a tenable solution, but one could last no longer than several days aboard a float, since they carry no rations or water. In cold water, survival on such floats depends on donning a survival suit before the vessel is abandoned. There is *never* an ideal time or condition for abandoning ship. This should always be an act of last resort, for one's chances of survival diminish significantly with a rash and hasty decision to abandon ship when there is still hope to save the vessel.

WORKBOATS AND DINGHIES

Commercial vessels have boats; yachts have dinghies. Even the smallest commercial vessel needs a boat not less than 10 feet in length, and larger is always better. To carry any weight in anything other than calm water, the depth of hull must be at least 16 inches, which makes the small boats rather tubby-looking. The boats are carried on deck or, if the vessel is large enough, on the hatch top. On vessels 50 feet long or longer, carrying two boats has definite advantages: the larger is for heavy work such as landing small cargo and package freight; the smaller is more or less for pleasure and light duty as a tender, yet should be stout enough to be used for running out a kedge anchor or for other emergency duties.

The failing of most stock boats is that they will not stand up to normal abuse, let alone the demands placed on them in extraordinary situations. Many do not perform well or cannot be rowed at all, especially under adverse conditions. Therefore, if the builder is required to furnish a boat, he would do well to modify it as necessary to allow it to perform its prescribed duty. In a wooden hull, modifications would include the addition of chafing strips to the bottom for landing on rocky beaches, and reinforcement at the gunwales plus an adequate rubrail, since the boats must lie against concrete pilings, creosoted wharves, and the sides of the vessel in a seaway.

In a glass-reinforced plastic hull, builders are prone to leave the skeg hollow, so this should be filled. Gunwales are inadequate and must have extra-heavy rubrails installed. The bow is also usually inadequate and should be reinforced. Lifting eyes and their pads should be attached to extra-heavy reinforcements. If there is any flexing at all in the bottom, hat sections should be added to stiffen it. In an aluminum boat hull, the major requirement is a suitable fender rail. In all types of boats, some sort of drain must be added to keep the boat free of water in its normal stowed position, unless one is already built in.

When the owner furnishes the boat, the only work required of the builder is to make sure it fits into its chocks and can be secured with its gripes.

One can get these boats over the side or back on deck in many ways: with davits, the vessel's halyards, rail rollers, or cargo booms and cranes, to name but a few. The use of portable davits is generating renewed interest, since it allows hoisting on either side—for example, the side that will be most convenient for the next stop. Also, if necessary, davits allow carrying the boats swung out or centered on the railcap, keeping the decks clear for working the vessel and avoiding interference with a deck cargo.

The use of stern davits to carry a boat has the advantage of clearing the decks, but if the boat is longer than the breadth of the transom, it will be a nuisance when coming alongside wharves and other vessels. Boats hung in stern davits are difficult to launch and retrieve in a seaway unless there is an adequate number in the crew or a mechanical device for hoisting. The crewmember who hooks in the falls will often have to climb a rope ladder or go hand-over-hand up a manrope to gain the deck. A vessel with low freeboard aft is apt to have the boat filled in a following sea; therefore, the boat must have extra-large drains installed. With normal freeboard, the drain need only be adequate to keep the boat free of rainwater. In spite of these negative features, the stern remains the best place to carry a motorized yawl (push) boat on vessels having no auxiliary.

As mentioned, yachts have dinghies or tenders. They vary from handsome, usable rowing boats, sometimes also fitted for sailing or motoring, to atrocious-looking, flimsy, poorly designed, miserable rowing containers intended to service vessels worth 100 times their cost. Inflatables top the list of tender abominations, and the vast majority cannot be rowed in any but the calmest of seas. They mostly owe their popularity to the desire of their owners always to increase cabin space at the expense of usable, working deck room on a vessel already hopelessly cluttered with gadgets that could have been eliminated had someone remembered the existence of blocks with more than one sheave. In reality, there is seldom space left on the deck of a yacht for properly stowing a proper dinghy.

Since most sailing yachts have lifelines, their dinghies must be lifted over them using a halyard. Motor yachts, having more than adequate electric power, can use a motorized davit not only to lift the dinghy and its motor, but to set it in the chocks with a minimum of effort. Some of the smaller trawler-type yachts have a mast and boom suitable for accomplishing this task in port; this is also used as a steadying rig at sea. When a boom is used, guys will be needed to steady the dinghy in any sort of sea, so several pad eyes should be added to the railcap to facilitate this. Stern davits on yachts have become quite popular and are usually purchased from the manufacturer or from a marine supply house, ready to install. The best installation involves welding a foundation either on the deck or suspended over the stern to accommodate the davit base plate. The overhang of the davit should equal the amount necessary to clear the dinghy of any stern rails or rigging that would otherwise prevent it from being secured in the hoisted position.

Dinghies, and workboats too, must have their lifting eyes as high as possible, and not less than the height of the thwarts, or they will capsize when lifted. On some of the heavier yawlboats, the lifting eyes are fastened directly to the skeg and stemson, in which case a trunk is required to prevent capsizing. This is a poor solution because, during much of the initial lifting before the boat becomes vertical, it causes excessive friction and chafe on the falls.

LIFEBUOYS, VESTS, LINES, AND OTHER DEVICES

Ring lifebuoys of adequate diameter are difficult to stow on most small vessels in order to make them usable on a moment's notice. Also, they can, when accurately thrown, knock out the recipient. The most favored type—the so-called horseshoe ring—is the easiest to use in the water, is generally made of a soft material, and takes less space for proper stowage. There should be at least two of these rings sited close to the steering station. Sometimes they are fitted with parachute drogues to keep them from drifting out of reach. Personal experiments with these lead me to believe drogues are a useless added expense, and less effective than a length of line with several fishnet floats attached, which also aids in the retrieval of the person in the water. Horseshoe rings fit into stainless steel holders, which need only be bolted to the bulwarks or some other permanent structure. Hose clamps to hold them in place when fastened to rails will eliminate drilling any holes or otherwise marring the finish of the vessel.

There is no question that a flagstaff, independent of but attached to the horseshoe ring and weighted to keep it upright, is an excellent aid in locating the buoy; however, it often interferes with the running rigging and booms on sailing vessels. Such a device has small value at night, and the use of a searchlight to locate it makes everyone on deck night-blind when they most need to have acute vision. Therefore, the use of a water-activated light at night and a water-activated smoke bomb in daylight is the better solution. On some vessels, the master insists that each person on deck at night have a personal strobe light attached to his clothing or worn on a lanyard, especially in heavy weather.

Life preservers—PFDs, or personal flotation devices—are required on all vessels. It is proper to construct a rack or provide shelf space at each berth, so the person occupying that berth has a personal PFD that fits him properly. Coast Guard regulations require one for each person, plus 5 percent more when making an international voyage. Since most small vessels carry fewer than 20 persons, this means one for each berth plus one extra. If space permits, *several* extras should be stowed in a convenient locker, preferably below decks and readily available, since extra persons or guests are often aboard.

All vessels will at one time or another find use for temporary safety lines and lifelines. While these may be fastened to a variety of objects, the builder should add several pad eyes or belaying points with this specific use in mind.

Flares and flare guns should be sited in a proper locker, installed as near the steering station as possible. In the case of outside steering, they should be sited near a companionway, accessible from the deck but not exposed to the weather. Vessels fitted with booby hatches rather than sliding hatches can usually have the locker located within these hatches.

A portable waterproof emergency searchlight should be carried, and it should have at least two waterproof plug-ins on deck. One should be on the foredeck, while the other should be on the steering gear box if outside steering is used, or in the wheelhouse with enclosed steering.

Most of this chapter might seem redundant and obvious, and it will be to experienced designers and builders who have spent some time at sea. Many have *not* had sea experience, however, as becomes quite evident from the state in which some vessels are delivered to the owners. Attending to seemingly endless little details makes the difference between a proper seagoing vessel and a makeshift, jury-rigged vessel aboard which every task is a major undertaking rather than a simple routine.

5

$\triangledown$

BALLAST
AND
BALLASTING

There are several schools of thought on ballast and ballasting—when it should be added, where it should be placed, what materials to use, and how to install it. From a builder's viewpoint, the best time to install ballast is as soon as possible after the hull is welded up, since no precautions need then be taken to protect the finish of any portion of the vessel. From the owner's viewpoint, early ballasting is also the least expensive. From the architect's viewpoint, 60 percent of the ballast should be installed early and another 30 percent just prior to launching, when final estimated weight changes can be calculated. The final 10 percent is saved for trimming after launching. From the seaman's point of view, ballasting should wait until after launching, and then only enough ballast should be put aboard for the initial trial runs. Based on the trials, ballast would then be added to make the vessel "feel right." Despite the self-importance felt by the architect furnishing the design, the builder furnishing the finished product, and the owner furnishing the money necessary to accomplish the design and product, all these parties must ultimately defer to the seaman's feel for the vessel's behavior. There is no way to describe this instinct, and it cannot be taught. Most seamen have "it," some to a fine degree of sharpness. Basically, there is a certain correlation between the pitching and rolling of a vessel that is needed to make her comfortable in a seaway. The state of the sea and the easiness of the vessel are infinitely variable, so the more or less accepted rules devised by guessers and mathematicians become ludicrous at sea.

In all vessels, some areas must or at least should be preballasted, sealed, or both, as it would be next to impossible to do a proper job after launching. These include the areas under the engine and tanks. Most ballasting should take place after launching, however, when the guesswork about the vessel's light weight ceases; the exact line of flotation may then be determined by the draft marks fore and aft, and the actual light displacement—which is, of course, the launching weight—can be calculated. Knowing the true launching weight will help the designer refine his weight estimates for future vessels of similar hull configurations, and also help the builder estimate cost per pound or ton, man-hours to this stage of completion and, later, number of man-hours from launching to delivery for other vessels of similar type or weight. The designer also has the opportunity to ascertain the exact vertical center of gravity (VCG) through a simple inclining experiment, which he can then use for future reference.

TRIALS

How much ballast should be placed aboard for trial runs depends on whether or not speed trials were made a precondition of the contract for the purpose of determining a bonus. If speed trials are required, just enough will be used to keep the propeller immersed and the bow down far enough to provide reasonable visibility, with the minimum of fuel and water and no stores aboard. In most instances, the contract does not specify a speed trial, so half the fuel, half the water, and enough ballast to bring the vessel to 75 percent of the in-ballast waterline displacement are carried on the first trial runs. The trial location for the ballast in most hull forms will be within 22 percent of the designed waterline (DWL) either side of the designer's midship section. The ballast should be spread evenly.

During the initial trials, the wave height in a seaway need not exceed one-twentieth of the vessel's deck length, which is sufficient to get a feel of the pitching characteristics. A change in the motion is accomplished by moving a portion of the ballast from one end to the other, maintaining the 44 percent of DWL limit. A vessel that lifts quickly will need some ballast moved forward, and one that will not lift at all will need some ballast moved aft. Once the motion is more or less agreeable, some ballast from each end should be moved toward the center. Moving, say, 5 percent from each end makes quite a noticeable difference. The total move at one time should be about one-third the length from each end of the ballast toward the center. A change either for better or for worse will be immediately noticeable. After trying out several combinations to discover the best motion in a head sea, the vessel will then make a couple of runs in beam, quartering, and following seas, which will indicate any peculiarities of hull form affecting her steering and directional stability. Some vessels in the light condition prefer being down a bit by the head or stern, even though this induces a less-than-perfect pitching motion.

At the end of the trial runs, this amount of ballast is permanently set in place, after a check to make certain the cavities are free of dirt, chips, and other debris. Any further outfitting is completed, and the ship's stores are brought aboard and stowed in their proper places. All equipment is then brought aboard and stowed, leaving the vessel complete except for the crew's personal gear, voyage consumables for the vessel, and victuals. Again ballast is added, this time to bring the vessel to the ballast waterline, and trials are conducted in about the same sea state as before. The procedure for distributing the additional ballast remains the same as before, but this time there will be less shifting of ballast, for it can be laid down proportionately to the previous ballast. The final shifting for the correct feel will be minimal. The same captain must conduct all trials, since there is always a difference of opinion between any two captains.

In spite of what the inclining experiments indicated at the dock with the vessel in the static condition, her dynamic feel, which is determined now, is all that really matters. If the vessel feels tender, then more ballast may be needed or the existing ballast may have to be lowered. The latter course of action is difficult, since the vessel is now to her normal ballast line and one assumes that the existing ballast was sited as low as possible to begin with. The options are either to add more permanent ballast, lessening the deadweight cargo capacity of the vessel, or always to carry temporary ballast when light. In rare instances a vessel will be found too stiff, so that, while she still needs the amount of ballast already aboard, some of it must be raised to slow up her rolling period. A quick roller is hard on her gear and her crew.

Of the total amount of ballast required, it is wise to retain about 5 percent as portable ballast, so that it can be used for trimming during the life of the vessel. In this way, changes in operating areas can be accommodated with a minimum of effort. An example of this would be a vessel originally fitted out for deepwater anchorages requiring large quantities of chain, but later in life frequenting shoal water anchorages requiring a lesser amount of chain. Trimming ballast might then be moved into the forepeak to accommodate the decrease in weight forward.

Power vessels with their machinery well aft can often dispense with most or all permanent ballast and depend entirely on forepeak and afterpeak ballast tanks, should these be required. There are and have been sailing vessels that need no ballast when light and can, under reduced canvas, make coasting voyages in safety. It is doubtful, however, that many of them would undertake a long ocean voyage in the unballasted condition. Certain types of yachts, such as high-speed powerboats and multihulls, neither need nor carry ballast.

Yachts present a different problem than do commercial vessels in that the only variables they have to contend with are fuel, water, and stores. A vessel used only for weekends and vacations has the smallest ballasting changes with which to contend. Be they power or sail, however, yachts that undertake long voyages, the legs of which are measured in days and thousands of miles, do experience a considerable change in draft and trim from fully laden with stores, water, and fuel "on departure," to a light condition "on arrival." In these vessels the ballast is trimmed, if possible, so that as the vessel takes on or consumes fuel, water, and stores, her line of flotation remains parallel to her load waterline (LWL), and she does not trim drastically either by the head or by the stern.

Yachts will benefit from the same trial procedure used on commercial vessels, but the ballast added will immerse the vessel toward her LWL. With a yacht designed for ocean voyaging, it is best to fill the water and fuel tanks for the second trial and reduce the total ballast by allowing approximately one ton of weight per person making the voyage for food, stores, and personal effects. Therefore, the final ballast line may be several inches less than if the same vessel were to be used for coastwise cruising. It is always easier to add ballast than to have to remove it after it has been cemented in. In spite of all precautions, most vessels begin their voyages an inch or so below the LWL. However, the practice of continually raising the boottop to accommodate the added weights necessary for ocean voyaging should be avoided.

TYPES OF BALLAST

The common types of ballast are concrete, concrete and boiler punchings, pig and scrap iron laid in and covered with concrete, cast steel externally welded to the hull, cast lead carried internally, and lead ingots or bricks laid in and covered with concrete. For odd-shaped areas, lead is easily cast using a steel box of the desired shape. Ingots are convenient to handle in the 25- to 30-pound size, but from 50 pounds upward, they become awkward to handle and to site. Many vessels use paving blocks, bricks, beach stone, iron ore (washed), and kentledge, laid in the bilge either loose or stanchioned down, as a substitute for the heavier cemented-in-place materials. However, this allows dirt, water, fish scales and slime, and other contaminants to seep through, causing a foul-smelling bilge as well as preventing the circulation of air over the shell. The shell remains moist, which accelerates the deterioration of the coatings.

TABLE 10. Weights of Materials Used for Ballast

Material	Pounds per Cubic Foot
Lead, cast or bricks	710
Cast iron	450
Iron ore, limonite	237
Iron ore, magnetite	315
Iron slag	172
Lead ore, galena	465
Steel bars	490
Steel rounds	385
Steel balls	257
Steel boiler punchings	235
Bricks, common	120
Cement, sand, and stone	144
Stone, granite rocks	96
Sandstone	82
Riprap	80–105
Asphaltum	81
Pitch	69
Tar, bituminous	75

In sailing vessels there is little excuse for using any material other than lead, since it stows in the minimum space and has the most effective righting lever. The cost saving of using a less dense material is probably the poorest excuse normally offered, when one considers the number of gidgets and gadgets that owners purchase at great expense and that contribute nothing to the safety or seakindliness of the vessel. Some people seem to save with a spoon and spend with a shovel.

CEMENTING BALLAST IN

When an appreciable amount of ballast is needed in a small space, lead bricks are the best choice. These stow with the same density as cast lead (710 pounds per cubic foot). When bricks are used in a box keel, the keel sections are first cleaned of any dirt and trash, then coated with pure Portland cement and water mixed into a paste and thoroughly brushed over all metal with which the lead could come in contact. The lower course of lead bricks is laid in a grout of cement, sand, and water, mixed either from separately bagged materials or a premix available from a building supplies store. A space of not less than ¼ inch is left between the bricks and the steel floors or keel sides. The bricks are laid one on top of the other as closely as possible, without spaces, and the gap between the lead and the steel filled with grout. With all the ballast in, the top is capped with a firmer mix of the same materials, at least 1 inch thick but not more than 3 inches, unless there is a void or some fairing to do. After the cement has cured, it is coated with a suitable sealer and then painted.

The final cementing-in of ballast calls for extreme care to avoid any water traps that might restrict drainage to the sumps or let any bilge water collect at a low, unlimbered point. In power vessels utilizing a bar keel construction, it is usually necessary to carry the cement to either the first or second longitudinal, depending on deadrise. The best way to assure drainage outboard of these longitudinals is to fit either a plastic hose or tube through the limber cut in the longitudinal, directing any water to the sump. A limber chain should be used if there is much curvature in the hose or tube lead. If it is more or less straight, then a plumber's snake run through occasionally will suffice to keep the tube clear. The tubes are encased in concrete and decline toward the sump when the vessel is trimmed to an even keel. The deep floors must be cut or drilled for the tubes to allow the water to reach the sumps. Just prior to the final cementing, it is customary to drill the limbers, using a hole saw. When the top of the cement is of great breadth, it is best to cast along the centerline of the vessel a trough to catch the water and direct it to the sump. Any readers familiar with flat-bottom craft will appreciate the difficulty of draining those last few gallons from the bilge—just a little bilge water looks like the whole ocean sloshing about.

The concrete is troweled smooth using a steel trowel, but the initial work is done with a wooden float. The concrete should not be worked more than necessary, since overworking causes the cement to come to the surface, resulting in a weak mixture. In deep sections, a vibrator may be required to eliminate any air entrapments; this should always be used sparingly. Wet burlap laid for several days over the concrete will minimize the risk of cracking on the surface. Concrete requires 28 days for a total cure, so any painting of it should be delayed for at least that length of time.

Premixed cement is also used at the bottom of the chain locker, both to protect the steel in way of the chain and to prevent the chain from fouling in the low sections of the hull. A depth of 6 inches usually suffices on most vessels. Elsewhere, cement is used in shallow, inaccessible voids (rather than plating them over) and around rudderports to provide drainage.

The addition of cement to fill the voids in, say, a compartment filled with steel balls would add about 65 pounds per cubic foot.

The use of pitch for a covering in lieu of concrete is restricted mostly to pure sailing vessels, for in power and auxiliary sailing vessels fuel oil coming in contact with pitch would cause it to liquefy.

BALLAST RATIOS AND STOCK PLANS

Sailing yachtsmen and designers make much ado over the ballast/displacement ratio. Using this ratio to compare two vessels of identical hull form only indicates that, if there is a significant difference, the vessel with the higher ratio is either built lighter or has less joinerwork, less equipment, less water and fuel, a smaller engine, or lighter ground tackle. When comparing, say, two 40-foot vessels of *different* hull form, this ratio becomes an absolutely meaningless number, one that has been foisted on the public. In steel vessels, all metal located below the canoe body (that is, the bearding line) can be considered as ballast; therefore, the actual amount of ballast needed to bring the vessel down to her lines may be substantially less than the amount required to ballast the same design built in wood, aluminum alloys, or glass-reinforced plastics. In any case, the shape of the hull, more than anything else, determines the amount of ballast required for seakeeping, the carrying of the rig, and the comfort of the crew. No minimum ballast/displacement ratio exists below which *all* vessels may be unsafe—only a minimum below which *a particular vessel* may be unsafe.

Stock plans may or may not indicate the amount of ballast needed. If a quantity is given, the builder must use this figure with extreme caution. In all probability, it does specify the correct amount *if* the plans are faithfully followed, but plans are rarely faithfully followed, and common substitutions, such as teak in lieu of cedar or ¾-inch mahogany instead of ¼-inch pine for ceiling, increase joinerwork weights at an alarming rate. The same holds true when the size of the engine or tankage is increased. Therefore, the required amount of ballast is the difference between the weight of the vessel with all her gear and her displacement at the LWL (in the case of yachts) or the in-ballast WL (commercial vessels). The additional weight caused by changes from the plans may amount to as little as 10 percent or so when only a few changes are made. The worst case I can recall involved a vessel launched without ballast, fuel, or water, that displaced 60 percent more than her designed displacement ready for sea!

6

$\triangledown$

PAINTING
AND
FINISHING

Perfect corrosion protection is possible at only one time during the life of a steel vessel—when the vessel is under construction. The longevity of the vessel depends on the excellence of the steel preparation, the workmanship, the paint system, and the finish done in the builder's yard, coupled thereafter with normal good maintenance by her crew. A good builder never takes shortcuts in surface preparation or coating, because they will lead to grief in the long run.

As with all things pertaining to the sea, *the best is none too good.* Contrary to the standard misconception that the best is also the most expensive, it is possible to use a variety of inexpensive coatings over properly prepared steel and obtain longevity; however, even the most expensive coatings will always fail when applied over a poorly prepared surface. Furthermore, with sophisticated coatings, maintenance should be carefully considered, since such coatings may require special equipment and skills beyond those normally expected of a vessel's inventory or crew.

The ideal time to start preparing for a proper finish is right from the beginning of the steel order, by assuring adequate protection of the steel prior to its being used. It should be kept as dry as possible, and no water should be allowed to stand on the surfaces. The drops (cutoffs and scrap) should be treated just as carefully as the rest of the material. Until the vessel is launched there is no such thing as a junk pile, for the scrap pile never ceases to yield useful items. As construction progresses, slag, electrode stubs, flux, dirt, and water should be cleaned up as often as necessary, and thoroughly at least once a week. The more you protect the vessel and keep her clean, the easier you will find it to eliminate mistakes and oversights, and to begin the final finish.

RUST AND SCALE

Rusting and corrosion of steel is a chemical phenomenon and will not begin in pure distilled water nor in pure air, regardless of humidity. To start the rusting process, carbonic acid—which undistilled water and air both contain—must be present, but it is interesting to note that, once begun, the process will continue without further addition of the acid. The primary purpose of coatings is to prevent oxygen and water from coming in contact with the metal surface, since, once sealed, the surface cannot rust.

The process of rusting occurs gradually, starting with a brown bloom that thickens as more metal is affected, and later becoming a thin, hard scale. Each new layer of scale forms *under* the previous layer and adheres to it. In time, this becomes a thick, slatelike covering, in which each layer has a different color. The outer layer and all succeeding layers except the one adjacent to the metal have a different chemical composition; these layers are very distinct in a chip of rust. The layer next to the metal always has a percentage of carbonic acid in it, which must be reabsorbed from the previous layer of rust to continue the process. The scale so formed ranges from four to six times the thickness of the metal from which it is formed. Thus, for example, if the rust were ¼ inch thick, it would represent a loss of steel ⅟₁₆ inch or less in thickness.

Although rust is an oxide of iron, it is not the *only* kind of oxidized iron, and differs in particular from the distinct kind produced during the rolling process at the mill by the addition of water to the hot steel. This scale is hard, thin, tenacious, lustrous, and blue-black in color, and is known by the names of "mill scale," "black oxide," or, because of a radiating pattern it displays when fractured, "magnetic oxide." When intact, mill scale is impervious to water and protects the steel from corrosion. It is seldom a perfect coating, however, and, being brittle, fractures easily. When this occurs, the plate, being exposed to seawater, forms a galvanic couple (battery), and the scale causes a rapid corrosion of the steel. This important fact indicates that there is a need for *entire removal* of all scale on the exterior of the vessel, but that its presence is allowable on that part of the vessel's interior that is *not* in contact with seawater, meaning all areas outside the bilge, with some exceptions.

Steel plates can, of course, be descaled chemically, and this is done in several industrial processes in preparation for subsequent coatings. Galvanizing is one such process. In one method of descaling, each piece of steel is dipped in a dilute bath of hydrochloric acid, followed by an alkaline wash, and then scrubbed with mechanical wire brushes. Many people think that "triple-dipped hot galvanizing" means three times as much zinc is used. On the contrary, only so much molten metal can be attached to any surface by dipping, regardless of the number of dips, and the description merely means that after a hot acid bath to descale and remove other impurities, such as paint or oil, the steel is neutralized in yet another hot bath and then immersed in molten zinc. In cases such as the galvanizing of chain, it is also tumbled (rotated) so that the links will be completely covered but not fused together.

The old practice of letting a vessel rust during construction has been found uneconomical, especially in a salt-laden atmosphere, since severe pitting can occur over an unprotected surface. The practice of just wire-brushing the exterior has also almost ceased, since it has been determined that it is impossible to remove all the contaminants in the bottoms of the pits by this process. If these contaminants are not removed, they will permit corrosion to continue despite the perfection of the overcoats on the surrounding metal surface.

To paint over scale, flaking, and decomposed paint is a waste of time and money, although owners often do it to beautify a vessel for sale. This not only delays proper maintenance, but also allows the metal to continue corroding, thus becoming weaker.

All coatings will eventually decompose if not cared for, but the marine paint and chemical companies have developed coatings that will provide a long initial life and permit overcoating at a later period of the vessel's life. Many of these coatings will reactivate to some extent the previous intact coatings, which further extends the life of the coating system as a whole.

TRADITIONAL COATINGS

Some older paint systems manufactured by each yard to its own recipe required complete removal before overcoating. Some of these old recipes are nevertheless excellent, and I have included them at the end of this chapter, not as a substitute or a way to save a few pennies, but as a means of repair and maintenance in remote areas. I have repaired steel vessels in places where import duties came to 200 percent of the material invoice, and when packaging, shipping, bribes, and delivery costs were added, the final cost would have exceeded 20 times the original cost. Delivery delays of perhaps several months would have led to further expenses from the wages of the crew, loss of revenue, and many other interrelated factors. Given the circumstances, suitable substitute ingredients were obtained locally for one-half the original material cost of the commercial products. While perhaps not as good or convenient to use as the products available elsewhere, the substitutes were certainly suitable for the application and the situation.

Watertight coatings of Portland cement, tar and cement, bitumastic cement, or black varnish have been used since the beginning of metal ship- and boatbuilding. The bituminous cement is not affected by heat and is usually applied in coats ½ to 1 inch thick. It lasts the life of the vessel. Bitumastic solution is a bituminous material that is applied cold and may be used anywhere on the interior. Bitumastic enamel is a quick-setting material, applied hot (about 160 degrees Fahrenheit). Coal tar, tar varnish, and black varnish, all of which use asphalt and bituminous compounds, are impervious to water but soluble in many of the fuel oils and other petroleum distillates. The modern coal tar epoxy is also applied cold but gives a very hard surface. It is impervious to water and most chemicals.

Cementitious coatings or a wash of pure Portland cement may be applied to wet plating as well as to incompletely cleaned or descaled plates. While not impervious to water, such coatings are alkaline, and, since steel will not rust in lime water, work excellently in wet areas such as tanks, chain lockers, and bilges. The modern, specially formulated cementitious coatings have the advantage, since they need less frequent renewal than a cement wash. Portland cement is used in the ballast areas as mentioned in Chapter 5, and the steel and ballast it covers seldom give any trouble, although when you use cement, you must work it up the structurals so that no crevice is formed that will encourage the entrapment of water. Cement makes an excellent repair material, for it will set up underwater. Sharp, clean sand must be used for any strength; silica sand is best. Seawater may be used for mixing without any harmful effect on the strength of the mix, but it will retard the setting time fourfold. Cement will adhere to *any* clean surface.

PLANNING THE JOB

The logistics of finishing a vessel pose a difficult problem in that each area of the vessel, whether internal or external, has a different coating requirement. While one *can* get by using only one system, I have always felt that, after expending the time and effort to construct a vessel properly, one has no excuse for skimping on the finish.

Before any paint can be applied to the vessel, the steel must be prepared to accept the primer coat that provides the transition base for bonding the subsequent coats of paint. The preparation required will vary depending on the paint system to be used, but in general, a surface blasted to white metal is acceptable for all paint systems, and is the best surface to achieve. There are other acceptable surface preparations which, if properly done, will assure the longevity that should be expected of all steel vessels. The choice of preparation and paint systems mainly depends on the builder's preference, especially when it comes to the selection of companies with which he may wish to deal. The information presented in this chapter will allow you to form an opinion and make a judgment based on your particular requirements. To my knowledge, there are no disreputable paint manufacturers, and this includes the many I have not listed at the end of this chapter. There is no reason to suppose that you should avoid any company's products, and the companies not included in the list were omitted simply because I have never had the opportunity to use their products personally: to pretend or imply specific knowledge of a product or to accept only the manufacturer's say-so would be an inexcusable disservice to the reader. Always remember that the paint companies are competitive, and in order to appear as cost-effective as possible, they quote the minimum thickness required for a given amount of coverage. To use the minimum on new construction is poor economy. While there is no such thing as a perfect coating, one should at least strive for perfection.

INTERIOR SURFACE PREPARATION

The customary practice is to begin with the interior of the vessel, since it is the most complicated area and takes the most time. Once the interior finish has been completed, the joinerwork, engine installation, piping, and wiring can be done; meanwhile, the fitting out of all areas above the deck proceeds without interfering with the work below decks. The decision to proceed in this way has several ramifications as to the method of preparing the steel for finishing and the choice of a coating system.

When a coating system is to be confined to a vessel's interior, there are seven methods of preparing the steel, but only two of these will work with *any* paint system. Therefore, one must first select a system and then make sure that the metal surface preparation meets the conditions recommended by the paint manufacturer.

● *Prepriming the steel as delivered.* When the steel is delivered from the warehouse, it is first stored in a dry place, elevated above the ground and covered at all times with a tarp. Each piece is then removed for a prime coat and a shop or barrier coat of paint. The individual pieces are inspected for tight mill scale and wire-brushed where necessary to remove any loose scale or rust blooms. After brushing, each is painted with a vinyl wash-coat primer of a two-part mix, usually phosphoric acid and a zinc chromate paste, followed as soon as dry (about 30 minutes) with one vinyl barrier coat thinned to the maximum suggested by the manufacturer. The wash primer must be overcoated in less than eight hours and never left overnight.

Alkyd paints are also used as primers, usually in the form of red lead. The use of alkyds was one of the earlier methods developed; it permits weld-through capabilities if the first coat is well thinned, and a well-thinned first coat is all that is needed to serve as a barrier and protect against further rusting. When the paint is dry, the steel must be stored with good drainage but does not require complete isolation from the elements.

There are, in addition, specially formulated, weld-through primers that are compatible with most overcoating systems. Many of these can be applied to a clean, dry, and tight scale surface. Steel so-treated can be burned, and one can weld through the coatings, which do not contaminate the welds as would a vinyl or alkyd coating.

Regardless of the priming method chosen, after each compartment is completed, the welded, scuffed, and burned areas will be power wire-brushed, swept clean, and primed in the same manner as the original plate. Prepriming is suitable for vessels constructed either indoors or outdoors. The plates should be painted on both sides so that the builder has an option, when laying out more than one pattern on a given plate, always to have a painted surface on the inside of the vessel.

When the vinyl priming system is used on the interior, the prime coat should ultimately be overcoated with a complete, five-coat vinyl system, plus two or three coats of alkyd paint. The shop coat is not counted as a final coat. The life expectancy is between 15 and 20 years of service except in areas that are always wet or have high abrasion traffic. The primer on the exterior will always be blasted off for proper finishing. Amateur builders seldom use the prepriming system, as the amount of time required to prepare the steel detracts from the progress they hope to achieve each day.

● *Prepriming the steel as fabricated.* With the steel covered as recommended above, each piece is removed from the pile as needed for cutting and assembly, and the fabricated part is primed and painted. This method is practical for frames and other fabricated parts; when one is fitting longitudinals and plates, however, there is always a delay while the paint dries. Some builders will, in the case of the longitudinals, paint what they have fitted to the hull each evening, making this the last task of the day. If the vessel is not being constructed under cover, there is always a chance that rain will occur before the steel can be painted, which causes extra work to prepare the steel properly. When plating, this procedure may also be followed: after one plate is cut, it may be painted while the plate on the opposite side of the hull is being templated, cut, and painted. This is practical in a one- or two-man operation, but with a larger work force the plates never seem to dry thoroughly, so it can become messy. The life expectancy of a coating applied using this system correctly is the same as that for the first method. However, priming as you go seldom produces as thorough a job, and a loss of five years of life expectancy seems a reasonable figure for the average builder.

● *Wire-brushing the interior.* Using this method, one wire-brushes, primes, and overcoats the metal after completing all burning, welding and chipping. In other words, the steelwork is finished first. This method is acceptable provided the steel is stored as recommended above, the vessel is built under cover, the scale remains tight for the most part, and the surfaces have very little rust. Delaying the inevitable acceptable surface preparation ultimately necessitates a greater expenditure of labor, since portions of the hull will no longer be accessible to power wire brushes and will have to be done with hand wire brushes and sandpaper. On vessels built in the open, the whole process is too laborious to be considered, and any "perfection" in the result is, at best, more a matter of faith than of fact. The life expectancy of this coating system is 10 to 15 years on vessels built under cover, and up to 7 years on vessels built in the open.

● *Brush-blasting the completed interior.* Brush-blasting is a procedure that economically removes loose mill scale, rust, and paint, as well as all oil, grease, dirt, and other contaminants, but retains the tight mill scale, rust, and any well-adhering paint or coatings. The entire surface must be uniformly exposed in blasting to reveal numerous flecks of the base metal. This is a more expensive solution to a problem easily solved in the beginning, using the prepainted steel on a vessel built in the open. (Note, however, that if the workmen are careless and cause extensive damage to the prime coat, even a preprimed interior will require brush-blasting.) If, on the other hand, the vessel is built under cover, brush-blasting is often all that is required to prepare for coating systems such as water-based cementitious coatings, some coal tar epoxies, chlorinated rubber, and alkyds, and prepriming can be eliminated. In preparation for the vinyl and older, oil-based coatings, it is a more than acceptable treatment of the metal, since it provides a tooth for adhesion. Because there will be some spots of base metal showing, repriming of the entire surface will be required. The used material (sand) must always be swept and vacuumed from the area prior to applying any wash primer or the first coat of paint. When properly done, the life expectancy of the coating is up to 20 years, depending on the system used. Although some coal tar epoxies and chlorinated rubber coatings can be applied after brush-blasting, these coatings prefer a better surface preparation for maximum longevity.

● *Blasting to commercial standards.* Blasting to commercial standards requires a surface free of oil, grease, dirt, mill scale, rust, paint, and other foreign matter not embedded in pits. At least two-thirds of each square foot (67 percent) must have a uniform, grey-white color, free of all residues, while the remainder may have slight discolorations or staining. In badly pitted steel, this is about all that can be achieved; needless to say, discolorations or staining should never occur in new construction. Blasting to this grade is most commonly accepted in the repair of older vessels and in the refinishing of badly maintained hulls. Some builders, however, consider it more than adequate for the interior of the vessel, especially if the preapplied shop coat is incompatible with the final paint system they plan to use. The life expectancy with this blasting depends on the coatings used. In general, one can expect the oil, alkyd, bituminous, vinyl, chlorinated rubber, and cementitious coatings to exceed 20 years.

● *Near-white blasting.* Near-white blasting requires the surface to be free of all oil, grease, mill scale, dirt, rust, paint, and other foreign matter, with the exception of some slight streaks or discolorations. Should the surface be pitted, very slight residues are permitted at the bottoms of the pits, but at least 95 percent of each square foot of surface area must have a uniform, grey-white appearance. The remaining 5 percent may have slight discolorations or streaks. In some instances a paint company will require a surface 99 percent clear of all defects. The life expectancy equals that of the coating, and when one has achieved this state of perfection, the use of coatings that are suited to each compartmental need becomes possible. This further reduces the maintenance and increases the life of each coating.

● *White blasting.* Blasting to white metal finish can seldom be achieved except in new construction, but in new construction there is no reason to stop short of this, since the base metal surface will have no contaminants of any sort and the adhesion of *any* paint system will be at its best. Coating failure can only result from improper mixing of the coatings, slovenly workmanship, or both. When a white metal finish is obtained, there are no imperfections. Many may argue that this is an extreme for which there is no need to strive, but the additional cost of white blasting in new construction amounts to perhaps a maximum of 2 percent more in labor and materials than near-white blasting requires.

The notion that any coating can be perfect is a delusion, because it is unreasonable to assume that in thousands of square feet of surface there will never be a pinhole in the coatings that leaves the steel unprotected. However minute the area, deterioration will commence in that spot. The number of pinholes can be lessened by changing the color or shade of each overcoat required to build the coatings to the manufacturer's recommended mil thickness. "Holidays" will then stand out. Note again that this thickness is usually the minimum that can be accepted. In new construction, or at any other time in the life of the vessel, there is no excuse for just achieving the minimum. The cost of paint applied during construction is insignificant compared to the labor of correcting an inadequate job at a later date.

I have used all of the seven methods listed. For reasons of economy, I often chose the first method, and found no fault with it provided that certain areas (discussed later) subsequently received extra treatment for their special needs.

EXTERIOR PREPARATION

The exterior of a new vessel is limited to only two possible preparations: the new surface must be blasted to "near-white" or "white" metal condition. When building in the open, some builders permit the exterior of the hull to weather (rust) in order to break down the mill scale and make the surface easier to sandblast. While the mill scale will indeed break down, one is always tracking rust into the interior, which in turn causes other problems. If the whole vessel is to be blasted inside and outside, doing all of the prime painting at one time, then most of the objections disappear, provided that the surface of the metal has not become pitted in the process of rusting, and the time element has not been long enough to allow the weathering to weaken the steel. This subject is covered in detail elsewhere in this chapter.

The more usual practice is to apply a white latex field coat to the plates as they are fitted to the hull, to delay rusting. In the tropics this will also cool down the hull for those working inside. The seams and butts are not painted until the welding is complete. The latex paint is easily removed during the blasting.

BLASTING METHODS

Vessels are seldom blasted in an enclosed area such as the building shed, because the dust goes into every nook and cranny. This means that the vessel must be moved from the building ways for blasting and priming, and then moved back either to the original building site or to another site, releveled, and blocked before one can proceed with the joinerwork. When the vessel is constructed outdoors, all support structure must be cut away and temporary or portable supports substituted, because the blasting of the exterior must be absolutely complete. These considerations often make prepriming during construction economical, to eliminate the need to sandblast the interior. The builder may then elect to finish off the interior, afterwards masking off all ports and other openings, blasting, and finally painting the exterior as one of the last tasks before launching. In this way, the vessel remains level, and the building shores need not be cut away.

Sandblasting at its best is a dirty, nasty job. If one could do all the blasting and then clean up and dust everything off before priming, it might not be so bad. However, it is necessary to stop frequently, clean up the freshly blasted area, and prime that amount of surface before continuing. How long one can safely blast before having to overcoat the surface depends on the weather, which never seems to cooperate. Blasting cannot commence before the dew has dried. On humid days, two to four hours will be the maximum; on dry days, the time may be doubled. In any case, the whole area that has been blasted must be overcoated before any rust bloom appears, and all work primed the same day, for it cannot wait until the next. There are no ifs, ands, or buts in this instance. For someone to touch a freshly blasted surface with his bare hands was, in my yard, a sure way for him to find out if there really is life after death.

When blasting, select an area of several square feet and outline it with a thin blasted line. Subdivide the enclosed area into smaller squares and blast each square clean before moving on to the next. If there is a lot of loose scale on the surface, it is best to brush-blast this off so as not to have these heavy contaminants blown back onto a clean surface. Needless to say, one should not skip around, but should systematically work from the clean surface toward the scale surface and from top to bottom. If at all possible, it is better to keep the blasting in a downwind direction. Blast the decks after the topsides; otherwise, contaminants will settle on the freshly primed surface with the probability that they will later bleed through the deck overcoatings.

One material used for blasting is clean, sharp, dry silica sand. Most of the smaller building yards find bagged sand (100 pounds per bag) the easiest to store and handle. The larger yards often purchase bulk sand in hoppers containing 5 to 10 tons of dry sand or, when sandblasting becomes almost a continuous yard activity, have their own permanently sited hoppers of up to 50 tons capacity. Bulk storage saves much time in loading the blasting machines, because one man may service several operators at once, plus an automated blasting unit.

Opinion varies as to how fine the sand should be. It is available in several grades, determined by what will pass through a given screen mesh. In some areas, sand is designated by numbers; in others, it is graded as fine, medium, or coarse. Some builders prefer to use coarse sand, which sometimes can be collected and reused once more. Used sand, however, contains the contaminants from the first shooting, and under the pressures used in blasting can become embedded in the surface. The coarse grade of sand makes deep, craterlike indentations on the steel surface. I prefer the fine sand, because it yields a finely textured surface, giving a satiny appearance. A rough surface permits embedding the primer coat deeply into the surface, but the adhesion is not as complete as when the fine sand is used. The comparison may be likened to painting a gravel versus a cement patio. In reality, neither grade of sand leaves pronounced craters, but the finer sand leaves substantially more indentations per square inch. The primers are quite thin, and some have a waterlike viscosity. The finer the surface tooth, the better adhesion these primers will have. With a higher-viscosity primer, however, a coarse surface is acceptable, since there is less likelihood of the primer bridging the craters.

Black Diamond and shotblasting are other blasting methods employed in larger yards in addition to the normal sandblasting. The Black Diamond material is fly ash, which is generally not reusable. The shot consists of steel balls that can be reclaimed and reused after being vibrated clean of all debris.

When the blasting and priming are to be subcontracted, then the whole vessel, inside (if necessary) and outside, should in all probability be done at the same time for the sake of economy. Blasting contractors are usually well equipped, in some instances even to using air-conditioned suits that permit the operators to stay comfortable and thus, one hopes, to do a better job. Contractors remove spent material with large suction hoses instead of the shovels, brooms, and small shop vacuums used in most yards, so the time element, too, favors this approach. The builder must still exercise strict supervision. Multi-primers are seldom feasible in subcontracted work, due to the cost of masking, stopping and starting, and cleaning the equipment. The need for some later spot-blasting is not eliminated, but this can be done with very small equipment.

If the builder has his own blasting equipment and needs only to rent a compressor, and if the interior must be blasted, there are certain advantages in doing the blasting and priming in two or more stages. Different primers to suit particular locations can be used as needed, and blasting can begin much sooner, allowing the builder to close off the completed areas and start the interior joinerwork. The exterior blasting may then be postponed to the last minute, when all welding and burning have been completed, assuring a smooth, continuous application of the primers and the overcoats.

Should the builder own the compressor, ideal blasting arrangements are possible. That is, he can blast when necessary and prime all the metal with a weld-through primer as he goes along, so that all he need do later is to spot blast in the way of cuts and welds.

ECONOMICS OF BLASTING VERSUS PRECONDITIONED STEEL

The Wheelabrator process, where all the material arrives at the yard preblasted or abraded and primed, is the best of all solutions. Unfortunately, it is not always available at a reasonable cost to the builder, since transportation costs, added to the charges for the service, can be crippling. As with the yard-owned compressor/blasting capabilities, only the cuts and welds will need spot-blasting after the vessel is finished. At worst, only a light brush-blast will be needed on the shop primer prior to overcoating.

On the U.S. West Coast, preblasted and primed Japanese steel is available as a stock item. The American steel companies seem to feel that if anyone wants their steel, he will take what they give him. The glass, resin, and paint manufacturers have spent millions of dollars in advertising and adapting their products for the builders and owners of small fiberglass (GRP) vessels. Had the steel, aluminum, and wood manufacturing companies spent even a third as much, there would have been a building surge in these materials, because metal- and woodworkers in the marine industry were numerous after World War II. Having dealt with the wood and metal industries most of my life, and having worked for one of the metal manufacturers, I attribute their neglect to greed more than anything else. They took the attitude that, since the volume was not there, they should concentrate only on high-volume orders, discontinue all specialty items with low volume, and let everyone adapt to them—never adapt themselves to the consumer.

I mention this because the cost of sandblasting is a major expenditure in the construction of a steel vessel. One bag of sand delivered to the builder will blast approximately 20 square feet of mild steel and 10 square feet of high-strength, low-alloy steel such as A 242. When sandblasting pipe railings, allowing 5 square feet per bag is generous. If there are numerous shapes, as on the interior of the hull, one should not expect to blast more than 50 percent of the area per bag that can be blasted on the external shell plating. These figures apply to clean steel with the mill scale intact. For painted or heavily rusted surfaces, and for those covered with oil, grease, and dirt, the number of square feet blasted per bag of sand drops drastically. With a ⅛-inch nozzle, sand consumption is about two bags per hour; with a ¼-inch nozzle, it increases to about seven bags per hour. (By the way, the larger nozzle is not always the best, since it usually consumes more sand per square foot when blasting shapes, pipe, and other such surfaces than does the smaller nozzle.) Blasting pressure should be 125 pounds per square inch (psi) at 125 cubic feet per minute (cfm). Assuming ideal coverage with the ⅛-inch nozzle, sand at $4.00 per bag, two men at $12.00 per hour, compressor rental at $80.00 per day ($10.00 per hour), and 2 gallons of gasoline per hour at $1.00 per gallon, the cost is then $1.10 per square foot. This makes no allowance for cleanup time, the cost of the primer, or the labor to apply it. Thus, preprimed steel could cost several cents per pound more than plain steel and still be economical for the builder of steel vessels. Preblasting and priming steel on a continuous basis costs the mill but a fraction of what it would cost a builder to blast and prime each piece of steel individually.

FLAME SPRAYED COATINGS

Galvanic action (electrolysis) can either increase or decrease the corrosion rate in a steel vessel. When a piece of iron and a piece of copper are immersed in a weak acid solution *without any metallic contact*, both will corrode. Should they touch or be placed in metallic contact with each other, however remote, a galvanic current will immediately be established between them. The copper will cease to corrode while the iron will corrode at a faster rate. Therefore, the immunity of copper from corrosion is obtained at the expense of the rapid corrosion of the iron. The result is, in essence, a voltaic cell (electric battery). The corroding metal is termed *electropositive*, and the immune metal is termed *electronegative*. When zinc is substituted for copper, the conditions are reversed, since the zinc is electropositive to the iron. Table 7 of Volume 1 lists the approximate potential in volts between the various metals in seawater at a velocity of 4½ to 7½ knots. Rust is electronegative to steel and thus will corrode the steel more quickly than if the plate were rustfree. This is true even when the plate is *not* immersed in seawater, since the rust scale is hygroscopic and therefore absorbs moisture.

Molten zinc and aluminum, flame sprayed to a sandblasted surface of near-white to white condition, have been used for many years, and at one time this was the best solution for the protection of steel. Of the two metals, aluminum was and still is the better coating in a marine environment; however, either of them poses one or more problems. In the case of zinc, pinholes through the coatings will allow salt water and, in bottom paints, other contaminants such as cuprous oxide and mercury to seep through to the zinc. The zinc then valiantly starts to protect the steel by disintegrating. As the zinc dissipates, exposing the steel, more zinc from the surrounding area comes to the rescue, but, being unable to reach the area, contents itself with protecting whatever it is already covering. The pinhole does not get bigger, but the dissolved zinc now forms crystals that absorb more water and contaminants, eventually forming a blister. The blister enlarges, contaminating more of the substrata, and so on. These blisters can range in size from as small as a pinhead to several feet in diameter. Corrosive salts build up within the blister, and then the steel begins to pit; thus, the savior of the steel becomes the destroyer of the steel. Aluminum, on the other hand, is galvanically more noble than zinc, and pinholes through it to the steel have a lesser voltage differential. The aluminum sacrifices itself at a steady rate, unlike the kamikaze zinc. If the coating system has a pinhole, the blister forms just as it did with zinc, but with a different chemistry of corrosion salts in the blister. (This same principle also applies to improperly welded clad steels. Once joint contamination occurs, undercutting and a separation in the bond of the cladding with the steel will follow. With proper welding, however, this will not occur and the steel and its cladding material will live happily together, never realizing that one could destroy the other.)

In either case, the chemistry of the salts contained within the blister is altered by further chemical linkage to whatever type of bottom paint is used. In effect, the hull has been told, "You are not steel; you are zinc or aluminum." All well and good, but many builders and overzealous paint salesmen do not recognize this, and continue to treat the vessel as a steel hull. Paints compatible with a steel surface more often than not are incompatible with a zinc or aluminum surface. Why, after all the research the paint manufacturers have done to eliminate the need for flame spraying, will builders and owners not accept that it is possible today to coat the entire surface of a steel vessel with inert materials? These materials not only bond to the steel but also form the required barrier coat, further reducing the hazard of disintegration of the whole coating system should a pinhole develop.

EXTERIOR COATING SYSTEMS

As previously mentioned, priming of the vessel is done as one progresses with the sandblasting in order to assure that no rust bloom will be trapped under the primers. Contrary to the best way of priming the interior, the best procedure on the exterior is always to use the same primer throughout, since doing otherwise requires extensive masking-off of the areas intended for different primers. I prefer to use the same primer required on the bottom of the hull for the whole exterior and then, if and where necessary, use a conversion coat that will permit a different paint system to be used with that primer. The various types of bottom coatings will therefore be discussed first. Note that compatibility with a primer is often only a one-way option. Coal tar epoxy will accept itself, vinyl acrylic, catalyzed epoxy, or chlorinated rubber as an overcoat; however, one cannot apply coal tar epoxy over chlorinated rubber, vinyls, or alkyds. To convert from a coal tar primer to, say, alkyd topside and deck coatings, one would need to overcoat first with a noncatalyzed epoxy primer, a vinyl acrylic, or a chlorinated rubber. Had the primers been vinyl or chlorinated rubber to begin with, then alkyds could be applied directly without an intermediate step.

Some primers, such as the wash-coat vinyls, will require an almost immediate second coating of anticorrosive paint to protect the primer, whereas many epoxy primers do not have to be overcoated for several months after application. Some epoxy primers are so formulated that, though dry to the touch, they will not fully cure until the next coat of epoxy is applied.

● *Vinyl pitch.* This system requires a minimum of a commercial blast, one prime coat, at least three anticorrosive coats, and two antifouling coats. The material may be applied with a brush or roller, or sprayed on. The anticorrosive coats have a life expectancy of three to five years before new coats must be added. The antifouling will repel barnacles up to three years and grass up to two years. The material will fail in the anticorrosive coat or by blistering of the antifouling coat. It is not difficult to clean. In spite of the alleged longevity, one overcoat every other year is normally applied.

● *Vinyls.* These require a minimum of near-white blasting, one prime coat, at least four (and preferably five) anticorrosive coats, and two antifouling coats. The material may be applied by brush or roller, or sprayed on. The anticorrosive coats have a life expectancy of three to five years. The antifouling will repel barnacles up to three years and grass up to two years. The material will fail by blistering and peeling. It is easy to clean. One additional antifouling coat should be applied on a 12- to 24-month basis, in spite of the supposed longevity.

● *Bituminous metallic pigmented coatings.* These are applied over a commercial blast, one primer coat, and three anticorrosive coats, followed by two antifouling coats. The life expectancy of the anticorrosive coats is two to three years, but the antifouling coats have a maximum life expectancy of one year. These coatings fail through peeling, undercutting, and blistering. Due to the cost of hauling coupled with a short life, it is seldom economical to use this paint system. The need to blast back to the metal is frequent, due to failure of the system, the subsequent repairs to the failed area, and the requirement of full overcoating on at least a yearly basis. It has the advantage of allowing almost immediate immersion after the antifouling is applied, and so lends itself to use on vessels requiring a quick turnaround or those that can be beached or careened during one tide.

● *Coal tar epoxy and epoxy systems.* These require, at the minimum, a near-white blasted surface. The primer is often dispensed with when using the coal tar epoxy, and the anticorrosive coats are applied directly to the sandblasted surface. The minimum number of coats required is two, but most builders prefer at least three. At least two antifouling coats are required. The life expectancy of the anticorrosive coats is from three to five years. The antifouling will repel barnacles up to three years and grass up to two years. The bottom is normally recoated every two years and usually requires a light brush-blast to give enough tooth for the next antifouling coat. Many small repair yards are prohibited from blasting due to its environmental impact, so some builders must, after the application of the anticorrosive coats, convert the system over to a vinyl or chlorinated rubber antifouling system, which permits the vessel to utilize any repair yard and is also much easier to work with over the long term. Epoxies can be applied by brush, roller, or spraying. Time and temperature are critical in the application, and one must not let them overcure, because brush-blasting or a solvent wash will then be necessary to retooth or reactivate the surface.

- *Chlorinated rubber.* This system requires a near-white blasting, three to four coats of anticorrosive, and two initial coats of antifouling as a minimum. The material may be applied with a brush, roller, or airless spray. The life of the undercoating is three to four years. The antifouling will repel barnacles up to three years and grass up to two years. It is normally recoated with one antifouling coat every 18 months or two coats every 2½ years. Today's cost of hauling warrants the longest period of time between haulouts, making the application of two coats of antifouling economical. Chlorinated rubber is an antifouling material in itself; when inorgano-tins are added, it is known as an "elastomeric matrix system."

- *Glassflake and polyester or epoxy resin.* This system requires a near-white blast, and it is applied by spraying. Builders frequently employ subcontractors specializing in its application. The life expectancy of the coating is at least six years at a 30-mil thickness, and rumored to be as much as 15 years using a 60-mil thickness. The coating is heavy, weighing about 0.375 pound per square foot at 30 mils and 0.75 pound per square foot at 60 mils. Unless this material can be carried above the waterline to a protective molding or to the underside of the deck-edge pipe, the chances of failure seem to increase to an unacceptable level, due to possible damage to a different topside coating that, when fractured, would permit corrosion to seep in and undercut the edge of the glassflake coating, causing it to separate from the metal. The overcoating for antifouling is the same as for epoxies when the glass is applied with epoxy binders. Chlorinated rubber and vinyl are used with polyester binders. Due to the added weight, I have had but one experience with the glassflake system, but found the application satisfactory.

When stating the life expectancy of the material, one assumes that the surface coating will begin to deteriorate at or near the stated time. If the coating is not attended to at that time, the underlying coats will then be exposed and begin to deteriorate. Since deterioration acts progressively, it is possible through neglect to destroy the whole paint system down to the surface of the steel. To prevent this, the overcoating system must be renewed at given intervals. When done over a sound surface, renewal will prolong the life of the substrata for an almost indefinite time. Any entrapment of rust or scale under a coating system can be considered sound only when the coating system itself has not been breached, since a breach will permit air, water, or contaminants to activate the rusting process. The importance of not applying extra-thick coatings of any material must be stressed, since a too-thick coating is a sure way to cause a paint failure. Even the high-build and mastic coatings impose limits on the film thickness that may be applied at one time, and one should never exceed them. The principal enemies of any coating system are sunlight (ultraviolet rays) and seawater. Lesser but still important foes include rain, sleet, snow, ice, oil, industrial pollutants, abrasion, animal feces, fish guts, and decaying flora.

BOTTOM PAINTS

One wonders why, with all the technical breakthroughs in chemistry, extremely long-lived paints have not become available. In fact, some of these paints *are* now available to the marine industry. Some bottom paints will last, barnacle-, grass-, and algae-free, for up to five years. Yes, they are expensive; however, net expenses must be computed not by the gallon of material, but by the cost of application, the cost of hauling, the loss of revenue from time spent in the yard, and the cost of not preserving the vessel. For the owner of a small steel commercial vessel or yacht, these factors may or may not be decisive. On the other hand, a concern of the paint companies is that many of the exotic coatings require special equipment; are dangerous to apply, because the flash point of the material is just below its explosive point; pose a real health hazard in the raw state; are for professional use only; and are not economical to package in pints, quarts, and gallons.

Better bottom paints are available everywhere except in the United States, since the Environmental Protection Agency apparently feels that it must preserve and protect the barnacles, algae, grass, and sponges rather than the vessels. Not being a moneymaking organization, the EPA ignores the necessity for the small entrepreneur to preserve his investment and make a profit. A case in point is the self-polishing copolymer bottom paint formulated and sold in England as well as other parts of the world by a major international paint company. After using two coats of this paint on the bottom of my vessel, I found her completely free of all growth, slime, and other contaminants, including a crude oil spill, after more than 25 months in the tropics. This was 16 months longer than I had previously gone with a top-of-the-line, U.S.–manufactured noncopolymer paint, which was so foul with grass after only four months that weekly scrubbings were necessary; after six months, pin barnacles needed to be removed on each cleaning. In the U.S., the manufacturer of the copolymer paint I tried cannot sell it as formulated abroad, and therefore markets a related product under a similar name. The two products are reputedly equally effective, and I must say that the American version, although it collects moss, grass, and algae after just four months of use, does repel barnacles for about two years. However, *any* growth on the bottom of a sailing vessel slows it down, so one must scrub the bottom before each passage.

● *Early bottom paint systems.* These contained primarily cuprous oxide (one-fifth to one-quarter by weight), mercury, or copper flakes. Today, mercury is not used in the U.S., so chlorine, tin, and cuprous oxide are the basis for most domestic antifouling paints. Some people have mixed in home additives, such as red pepper and bitter herbs, with the hope of prolonging the antifouling properties. I doubt that any of these additives do much, if anything, to extend the life of the paint.

● *Inorganic zinc coatings.* These have been around for many years and are extensively used on the topsides and decks. When used above the waterline and properly overcoated, the inorganic zinc coatings are considered lifetime primers, but are not a weld-through type of coating. These coatings, when scratched, have a self-healing quality that allows them to bridge gaps. Their principal use is in severe operating conditions, such as on oil rigs, crew and supply vessels, and barges. If applied too thickly, these coatings show a tendency toward surface cracking (alligatoring), so they are best applied with airless spray. In my opinion, inorganic zincs should never be applied below the waterline, since one is then treating a zinc hull rather than a steel hull. In recent years, however, chemical cross-linking of the zinc paints with organo-tins has produced a long-lived, anticorrosive, antifouling system reputed to last at least 48 months. I should stress again that the various paint companies are developing better products on a continuous basis. They do not have the stagnant, "it's-good-enough" attitude that prevails in many other industries.

● *Self-polishing copolymer bottom paints.* First formulated in the 1970s, these are composed of cuprous oxide and tributylin oxide bound into the polymer; this combination is hydrolyzed, releasing the antifoulants. Depending on the formulation of the paint, the desired leaching rate can be controlled when the vessel is moored or when she is maintaining her average speed underway. For sailing vessels, with their low speeds, the sloughing must occur sooner than for, say, motor fishing vessels operating at 12 knots. On high-speed vessels such as interisland passenger ferries operating at upward of 30 knots, this type of paint seems to have little advantage.

BOOTTOPS AND TOPSIDES

The boottop on most small commercial vessels is the bottom paint carried up to the load waterline, or halfway from the light waterline to the load waterline. On yachts, it is often a separate stripe of a contrasting color, sometimes of the same type of paint used on the bottom, or a different but compatible hard bottom paint that can be scrubbed. The usual practice, however, is to use the same type of paint as applied to the topsides, since antifouling paints of the sloughing type will streak the topsides if adjacent to them. Separating this type of antifouling from the topsides with a boot stripe will confine most of the streaking to the bottom. If continual scrubbing of the bottom is a necessity, a hard antifouling paint is recommended.

Topsides permit great latitude in the selection of coatings, but one should base his selection for a given vessel on certain practical considerations: the waters in which the vessel will sail, port conditions, frequency of hauling, hauling facilities, interim maintenance, repairability, ease of overcoating, and availability of the particular material.

● *Latex coatings.* The easiest to use, since the cleanup requires just soap and water, these coatings can be applied by brush, roller, or spraying over most primers, either directly or by using a conversion coat. For marine use, they should be of the acrylic type. Acrylic latexes come either in a flat or gloss finish; in a marine atmosphere, the gloss retention of the latter is only about six months. The ease of application and sanding allows one to do the painting from a raft or boat, and the rapid drying allows the vessel to go back into service immediately. The system fails through erosion, and a vessel sailed hard will probably need to repaint after each voyage. Repairs are easily made, and one can feather the edges of the repaired area by sanding. The density of the material will often permit a change of color or hue with just one coat. To overcoat, a scrubbing with detergents and a thorough rinsing is all that is required by way of preparation. The question of whether or not a saltwater rinse would suffice is intriguing, but to my knowledge unresolved. Scrubbing and initial rinsing of any painted surface with salt water is normal, but is always followed by a freshwater rinse.

● *Alkyd coatings.* These are the next easiest to use, since they are thinned with mineral spirits, can be applied by brush, roller, or spraying, and allow straight-forward, inexpensive cleanup. Alkyds retain their color for 6 to 9 months and will begin to chalk in 12 to 18 months. If chalking is permissible, recoating could be delayed for, say, three years, provided the barrier coats are not breached. To overcoat, the surface needs only scrubbing with detergents, rinsing, light sanding, dusting, and wiping down with mineral spirits. This method assures a clean, oilfree surface for proper adhesion. The paint should not be applied in heavy coats, but rather in two or more thinner coats; too much thinning, however, will cause a loss of gloss (as will exposure to dew before the paint dries). On the topsides, failure of the coating will be due to erosion, and a vessel sailed hard will need repainting of the topsides at least once each year.

Alkyds can be applied over a complete alkyd, epoxy, chlorinated rubber, or vinyl barrier coat system, which assures continued protection of the steel as the surface alkyd coating erodes. Another advantage of alkyd paint is that one can easily touch it up. The edges of any damaged area can be feathered, so the only telltale sign of repair will be a higher gloss on the repaired area compared to the remaining surface, which the next complete painting will automatically remedy.

● *Vinyl coatings.* These can be continued directly from the sandblasted surface, starting with a vinyl wash primer. In new construction it is good practice to use a complete barrier coat system: five coats to protect the steel, followed by one or two finish color coats. Vinyls are thinned with ketones and aromatic solvents. If the material is applied in thick coats, it will have a tendency to entrap the solvents, which will cause blistering; therefore, it is always better to apply a greater number of thin coats than to apply fewer, thick coats. Vinyls can be applied by brush, roller, or spraying. They have good color retention for about two years, and will last three years without recoating. The vinyl coatings fail through blistering (which one can minimize by using thinner coats) and by peeling caused by an improper bond to the previous coats. Lack of bonding usually results from painting over a damp or dirty surface. When spraying this material, one must exercise care to avoid air and solvent entrapment. Also, if the paint is insufficiently thinned, it will become stringy and will not have the proper adhesion. Vinyls will not yield a high gloss, but their ease of cleaning more than compensates for this. Repairs are easy to make, and the areas to be repaired can be feathered by wiping solvents over the edges of the old material. To recoat, a good scrubbing with detergents and a freshwater rinse is all that is required.

● *Epoxy polyamide coatings.* These may be applied by brush, roller, or spraying over an epoxy primer. In new construction, it is usual to use several coats as a barrier protection for the steel before applying the finish color coats. The material is a two-part mix; therefore, it has a limited pot life and is sensitive to temperature. Epoxies are thinned with ketones and aromatics. When a coating has overcured, it is necessary to wipe the surface with a ketone to reactivate the previous coat so that a bond will occur with the subsequent coatings. The coatings last approximately four years, but the gloss fades in 6 to 9 months and becomes chalky within 12 months. Its abrasion resistance and ease of cleaning will sometimes offset the difficulties encountered in repairing the surface. Although epoxy sands well and the damaged areas can be feathered, the two-part mix with limited pot life makes repairs tedious, especially in the tropics, and the repaired area is always noticeable. Epoxy systems fail by embrittlement, for which there is no cure; by blistering, which can be avoided by working with a clean surface that has been reactivated or brush-blasted to obtain the required tooth; and by air entrapment, which should be avoided, especially when spraying. Many people consider chalking a failure, but this is only true when it reaches the advanced stages which, when further neglected, will permit contaminants to adhere to the surface, accelerating further damage.

● *Chlorinated rubber coatings.* These can be applied by brush, roller, or spraying over a primer of the same material, and can be thinned with aromatics. The coatings have a useful life of 2 to 2½ years and retain color and gloss for about one year. Chlorinated rubber is often incorporated with binders such as alkyd resin to increase the water and chemical resistance of the paint film. The coating will fail through blistering, which can be controlled by proper surface preparation and by avoiding air and solvent entrapment. Before recoating, the surface should be washed with detergents and rinsed with fresh water. In new construction, it is usual to use at least three coats plus the primer coat before applying the finish color coat. The coatings are easily repaired by solvent feathering of the old area. Wet sanding can always be used, but the material has a tendency to gum up dry sandpaper.

● *Catalyzed polyurethane coatings.* These may be applied by brush or spraying. A catalyzed polyurethane coating is a two-part system and so has a limited pot life, which high temperatures make even shorter. The thinners used are esters and aromatics. Brushing the material in hot weather is best done by two persons side by side, with the second person using a sponge rubber brush to smooth out the brush marks of the first. The material, even when well thinned, dries quickly, and any holidays must be skipped until the surface has cured. The overcoat time is critical, and if the surface overcures, it will be necessary to sand between coats. The life expectancy of these coatings is from four to five years, and they begin to lose their gloss after three years. The gloss can be brought back with polishing wax, which will last about five months. To recoat the surface, a complete dewaxing, scrubbing with detergents, and sanding is necessary. Minor repairs are possible, but, as with epoxies, the repaired area is always noticeable. Failure of the coatings occurs by blistering and peeling, which may, as with other coatings, be avoided. The surface of the coatings is very hard and the resistance to abrasion is excellent. When polyurethanes are dinged—with the flukes of the anchor, for example—the material has a tendency to chip, especially when the undercoats are also brittle. To overcome this, most manufacturers of this paint have formulated undercoatings that are more flexible than the color coatings.

DECKS

Decks may be coated with the same paint system used on the topsides, but most yards will employ a different and less expensive system, since any deck coating is substantially less durable than a topsides coating and will not retain its gloss and color as well. In new construction, it is normal to prime and to use the same number of barrier coats to protect the steel as is used on the bottom, since decks receive the most wear and abuse. Then, if necessary, a tie coat is used to convert the barrier coats for alkyd paints, these being the easiest to use and to touch up.

The decks of most sailing vessels are treated to obtain a nonskid surface. The materials used for homemade nonskid are numerous and include sand, walnut shells (ground), pumice, micas, Epsom salts, aluminum powder, fly ash, and Portland cement. The nonskid may be premixed in the paint or sprinkled in the wet paint as it is applied. The latter method gives the most uniform appearance; the former uses the least amount of paint. Silica sand sprinkled in wet alkyd paint provides the least expensive and most satisfactory surface. This coating is allowed to dry, usually overnight, swept clean of loose sand, and then overcoated with two coats of thinned paint. The nonskid surface should not butt any framing, bulwarks, cabin sides, bitts, deck machinery or other fixtures, but should instead be separated by a narrow stripe free of sand. If the nonskid runs right up to these areas, then the paint, salt, and dirt will tend to build up, causing a premature failure of the coatings.

One should always repaint decks with well-thinned paint, for thick coatings will not dry properly and the heavy buildup will begin to crack (alligator) and become paint-sick. The only remedy for this condition is to sandblast back to bare metal and start over. The alkyd/sand–finished deck, applied over a complete epoxy, chlorinated rubber, or vinyl system, will last upward of 30 years with the steel in the same condition as when the vessel was built, provided that any damaged areas are properly repaired and the alkyds are always applied in thin coats and only when necessary.

As I have mentioned several times, decks in the tropics should be painted white, since white decks do not absorb heat as tinted or colored decks do. Gloss retention is next to impossible on any deck, since it must be frequently washed down, often with detergents, and has continual traffic over it.

Other deck coatings are used, some being troweled on and others being nonskid synthetic materials that are glued in place. Most of these are either too heavy or too expensive when compared to the materials mentioned previously. Some other deck coatings that can be applied by brush or roller eliminate the weight penalty, but unfortunately, few are available in white.

SUPERSTRUCTURE, TANKS, AND MISCELLANEOUS COATINGS

The superstructure will normally use the same coating material as the topsides. If any change in the coating system is contemplated, the usual choices are alkyds or latex for economy, or polyurethanes for longevity.

Fuel tanks should be sandblasted and epoxy coated. The theory that fuel oil will protect the tank is true, but it is not true that only fuel will find its way into the tank. Some water inevitably enters the tank due to condensation, bad fuel, or leaking fill pipes. Water settles to the bottom of the tank, causing corrosion, especially when combined with some of the impurities and contaminants that may be found in fuel purchased in certain areas.

Water tanks should be sandblasted and coated with a water-based cementitious coating and then a float coating that further protects the steel. Tanks I have built using only a float coat have provided 25 years of rustfree water. In new construction, however, a belt-and-suspenders approach is in order, so the double-coating of water tanks is the best way to go.

I have often used chocking material of a two-part epoxy mix manufactured by the Philadelphia Resins Corporation, both above and below decks, in lieu of wood, rubber, and other materials normally used to align or level the foundations for the main engine, auxiliary engine, and deck machinery. In twelve years of use on several vessels I have built, it has never to my knowledge deteriorated or failed.

Fairing compounds are used to create a smooth surface on hulls and to patch up nicks, scratches, and dents. Unfortunately, builders often use them in steel construction to cover up a surface that could have been smooth to begin with, given proper welding techniques, the proper framing, and proper lofting. Sometimes, in order to save steel weight, builders omit much of the framing, especially longitudinals, which will usually result in the need for fairing compounds. Fairing the entire hull involves a great deal of labor, and the end result, except when new, is far from satisfactory, since any severe impact or rust will dislodge the compounds. Extensive coverage commonly leads to local voids, which are then subject to a freeze/thaw cycle, causing large hunks of the material to pop off. Fairing compounds have one justifiable use: to repair damage to the finished coat after the vessel has been in service. When using filler, one should choose a formula compatible with the paint system.

By way of concluding this general discussion on materials and coatings, two other types of products that are commonly used aboard a vessel should be mentioned. Aluminum pigmented paints are available in several grades, defined by the temperatures they will tolerate. Generally they fall into two categories—those that will tolerate up to 400 degrees Fahrenheit, and those that will withstand temperatures between 500 and 1000 degrees Fahrenheit. The former are often used above the waterline and on exterior portions of the hull that have been coated with a bituminous material; used in these areas, they function as a non–bleed-through barrier. They are also used for slushing down galvanized rigging after it has weathered, and on galvanized chain, such as that used for the bobstay and bowsprit shrouds. The higher-temperature paint is used to coat old Charlie Nobles, exhaust pipes, boilers, and in some cases the entire engine room.

Cold galvanizing materials are used for touch-up of galvanized parts such as windlasses, turnbuckles, and shackles. They adhere well to previously galvanized surfaces and can be applied over bare steel after the surface has been thoroughly wire-brushed or sanded.

GROUNDING SYSTEMS AND ISOLATION OF DISSIMILAR METALS

Isolation of the hull from all other materials and fixtures can be accomplished with a minimum of expenditure. As we have already determined, in no instance should dissimilar metals come in contact with each other. The same holds true for electrics, which should never come in contact with or be grounded to a steel vessel. Isolation blocks should be used to separate the engine from the hull, the propeller shaft from the engine, any steady bearings for the shaft from the hull, and the stern bearing from the hull. Electrically, on a DC system, there is no reason ever to use the hull as a ground. Under no circumstances should the engine form part of the grounding system, since this practice destroys the engine. Even if no galvanic damage were to result, the engine always rests in a certain amount of oil, which causes an intermittent grounding and arcing. Grounding should lead to the negative pole of the battery, and all electrical leads should lead from the positive pole and return to the negative pole.

Alternators fitted to the main engine are grounded to it; thus, the engine must be isolated from the hull to prevent an electrical ground. Stern bearings are available with an inert phenolic or plastic shell encasing the bearing. Other bearings and assemblies may be mounted on phenolic blocks which, in turn, are bolted to the steel. One can purchase valves to the exterior of the hull, made of iron and internally coated with neoprene, which will cause no electrolytic problems. Some interaction will occur between the bronze propeller and a stainless steel shaft. They are close to each other on the electrolytic scale, however, so there is seldom accelerated corrosion, especially if the propeller is painted.

If a steel hull is permitted to become part of the electrolytic problem, the use of sacrificial zinc anodes is required. If, on the other hand, the hull is isolated from all machinery, electricity, and dissimilar metals, zincs can be safely eliminated. Following this practice, I have not used zincs on any vessel for over 30 years.

PAINTING TIPS

Painting the propeller is very common on auxiliary sailing vessels and commercial motor vessels. The propeller is sanded, degreased, coated with a strontium chromate primer, and then painted with organo-tin antifouling. Copolymer paint is acceptable if the vessel's engine is only occasionally used. My own vessel, *K'ung Fu-Tse*, gets two years of growthfree propeller use with this system.

Around the waterline, from the load waterline to 2 feet or so below the light waterline, it is customary to add a third coat of bottom paint with a complete two-coat system, because the velocity of the water in small vessels is greatest in this area. If only one bottom coat will be applied, then a second coat is always added at the waterline area.

New vessels are usually hauled for an inspection of the bottom not sooner than three months nor more than six months after launching. This gives the necessary early warning of a coatings failure. If the bottom proves in good condition, one additional freshening coat of antifouling is applied, and the vessel is not hauled again, barring severe grounding or other damage, until the elapsed time for the system nears. The best time to haul is just before the time when further delay would cause undue labor in cleaning, and always before barnacles begin to attach themselves.

There are several million tons of commercial steel vessels sailing the oceans of the world. The operators have a vested interest in reducing the cost of maintenance, the fuel consumption, and the passage time; similarly, the paint companies have an interest in selling their products by making them better. In new construction of small steel vessels, it therefore behooves one to use the commercial marine paints rather than those formulated primarily for fiberglass yachts, even though the latter coatings will work on steel hulls. The final and beautifying coats can be of a yacht grade of paint; in most instances, however, in order to achieve the high-grade mirror finish, a bit more preparation with the commercial formulations will be necessary. For example, on a 50-foot schooner, the amount of paint to preserve the metal properly (inside and out) before applying the finish color coats is between 125 and 150 gallons, depending on the system used. Buying paint in five-gallon pails will reduce the cost of painting, and a builder becomes a bit more generous with the paint this way than when it comes out of gallon or quart containers.

THE PINKY SCHOONER

From what has been discussed thus far in this chapter, it is apparent that a builder should preselect a paint system before starting construction, choosing one that applies to his particular situation. The Pinky, as we have discussed her throughout Volumes 1 and 2, would be built by one person, in the open, with the minimum of protection and equipment. This situation will apply to many one-time builders and many small boatyards. I suggest the following painting schedule and scheme as one solution I have used myself under similar circumstances. I do not claim it is the best or only method to follow, but I do know it will work. At no time during the construction of a vessel should the economics of the completed vessel be ignored, in terms either of time or money. The primary purpose of building a vessel is to get it in the water and use it, and it should therefore have a long life with the minimum of maintenance.

In all cases, one should devise a coating scheme and follow it. Also, in all instances, the manufacturer's field representatives should be consulted for any late developments that the company may have made. Many builders find using the products of just one manufacturer appealing and convenient. During the early years of steel boatbuilding, others, including myself, found it necessary to use the products of several manufacturers, since each independently made great strides in the development of primers, barrier coatings, and anticorrosive coatings, but not necessarily in all three types of coatings at the same time. One had to use the best product available from each company, as long as it was compatible with the best from another company. From this habit of picking and choosing we developed a finishing system we felt at ease with, a certain combination that achieved the desired result. Today, each company offers a complete and compatible system from bare steel to finished surface.

The system chosen for the Pinky requires that the metal as delivered be immediately cleaned and given a wash primer followed by a thinned coat of vinyl, since sandblasting of the interior is an expense to be avoided. During construction, the vessel is kept as clean as possible, and at least once a week all areas where welding and burning have destroyed the prime coat are chipped, wire-brushed, swept clean, and given a wash primer followed by a thinned coat of vinyl. The damaged areas on the exterior do not need this treatment, but to prevent rusting, a latex paint is used on the bare areas, for all of the exterior will eventually be sandblasted. The interiors of the water tanks and the tank tops are blasted just prior to closing, and two coats of cementitious paint are applied to all enclosed surfaces. These surfaces are further cleaned and touched up after welding on the top plate, using the clean-out openings for access. After all welding has been completed and the welds checked at night for any possible flaws (see Volume 1, Chapter 13), any remaining interior surface is primed, and then the entire inside of the vessel is painted with five additional coats of vinyl paint. The coats are alternated, with the first coat red, the second having aluminum pigment, the third red, and the fourth aluminum, ending up with a finished red color in the vinyl. This makes a total of seven coats of interior paint including the wash primer. Over the vinyl, two additional color coats of alkyd are applied in the cargo hold, the forepeak, and the lazarette. The chain locker uses cement on the bottom, and then a compatible, semihard microparaffin solvent. Because the cabins will be insulated, with all joints almost airtight, the vinyl affords a better surface in these areas, making the insulation easier to attach.

The joinerwork is completed prior to exterior blasting. All holes bored in the steel frames and other members for attachment of joinerwork are coated with a rust inhibitor prior to bedding and bolting the wood to them. The interior joinerwork, primarily of tongue-and-groove woods, is given three coats of clear tung oil sealer on all surfaces. The exposed surfaces are then varnished with clear spar varnish containing an ultraviolet shield, and wet-sanded between coats. Five coats of varnish are used in total, except for the cabin sole, the saloon table, the companionway ladders, and the cabin trunk sides, which receive eight coats. The settee and berth bottoms, having been part of the mold loft floor, already have one coat of flat white paint, so after a light sanding, a second coat of flat white is applied, followed by two coats of alkyd gloss white. The overhead is prepainted before being bolted in place, and thereafter requires no further attention. Since the cabintop is wood, no insulation is necessary.

Any exterior welds not already ground smooth are then attended to, and the hull supports that were welded to the shell plating during construction are removed (after fitting temporary supports and bracing the hull to prevent its falling over). For a short time, the hull will need to be jacked up enough to clear the keel supports for blasting and painting. The portlights are masked off with several layers of brown paper grocery bags or empty sand bags fitted to each port. The companionway doors are sealed, and likewise all vents and the mast collars. The steering gear and the windlass are covered, and all previously installed but removable deck fittings are removed.

The blasting commences at the railcap and proceeds down to the lower chine or thereabouts, since scaffolding is needed for these areas. The bowsprit is also blasted at this time. In the tropics, blasting is stopped frequently, especially during the rainy season, and the blasted area is primed. The blasting is then continued on the opposite side of the vessel to permit thorough drying of the primer. When the bulwarks and topsides have been completed, the bottom is blasted and primed. When the keel supports are reached, the vessel is jacked up and blocked to leave about 1 inch of clearance in the notches. When blasted and primed, the vessel is reseated in the notches.

The decks, along with the cabin trunks, are blasted last, care being exercised to preserve the masking and not damage the wooden cabintop. Once the decks have been primed, the whole vessel is inspected and any areas inadvertently blasted after priming are touched up. The sand has to be shoveled and swept off the decks. With sandblasting and priming complete, the masking can be removed and the interior of the vessel vacuumed to remove the dust that seeps in during sandblasting.

The chosen barrier coat system is coal tar epoxy for the entire exterior of the hull. The portions above the LWL receive a refined grade of epoxy to allow painting them a color other than black. After the required anticorrosive and barrier coat buildup, the topsides are given a tie coat and converted to alkyd high-gloss black enamel. Above the main deck pipe, including the bowsprit, the finished color for the bulwarks is alkyd gloss white enamel. The cabin trunks, insides of the bulwarks, and hatches will also have the same finish as the outsides of the bulwarks. The decks use the epoxy primer but will have all barrier and anticorrosive coats of chlorinated rubber, and will then be painted with white alkyd enamel. The deck is striped off 2 inches clear of the cabin trunks and other deck structures and even with the inboard edges of the bulwark stanchions. Small sections at a time are then painted with gloss white alkyd and sprinkled with silica sand. After drying, the loose sand is brushed off and the entire sanded area is given two coats of thinned, gloss white alkyd paint to achieve the required nonskid surface.

The cabintops, of marine plywood used in the mold loft floor, are fiberglassed (laid in epoxy resin), allowed to dry, then sanded smooth and given a fiberglass primer, allowing conversion to alkyd gloss white enamel. They are striped off and given the same nonskid sand finish as used on the decks. All exterior woodwork is primed and given three coats of alkyd gloss white enamel.

The bottom is given the required tie coat to permit shifting to a self-polishing copolymer. In new construction, a minimum of five coats of antifouling bottom paint is used, alternating colors with each coat except the last two coats.

The masts, being steel, are also sandblasted to white metal and primed, in preparation for a two-part polyurethane, which requires special primers. This yields the best surface for resisting abrasion from the sail lacings and the jaws of the gaffs and booms. The color will be buff to the hounds and then white to the cap. The interior of the mast will have a bituminous coating, since the masts are to be used as ventilators and cannot be sealed airtight. The paint is applied with a mop pulled through the mast, a sloppy but effective method. Booms, gaffs, and yards are of wood, and so are given one coat of flat white and three coats of alkyd gloss white enamel.

Black topsides are easy to match and do not show tire marks, creosote, and other substances. White decks and bulwarks keep the areas which must be walked or sat on cool. The buff-colored masts heat up, thereby causing better ventilation.

Upon completion of all the painting, the bowsprit shrouds, bobstay, netting, and other fittings removed for sandblasting and painting are refitted, and the hatch cover is secured. The vessel is now ready to be lowered into her launching cradle.

HOMEMADE PAINTS AND COATINGS

Paint is a mixture of pigments with a vehicle. The pigments are solids; the vehicle is a liquid. Many of the ingredients needed to make paint are already available in the galley or ship's stores. Others are readily available in most parts of the world. Note, however, that the following are emergency recipes to get the vessel to a shipyard for proper repair and coverage, and do not contain in their formulation any of the sophisticated chemistry used by the manufacturers of marine and industrial paints to yield a long life.

White lead, zinc oxide, sulfate of lead, or lithopone plus inert ingredients are used for white colors.

White lead can be made by adding vinegar over lead covered by tanbark. The resulting powder formed by the corroding lead is crushed, sifted through a screen, ground in water, and dried. The dried mixture is then ground in linseed oil or tung oil to form a paste. The process of obtaining pure white lead takes several months, and I mention it only to assure the reader that the mixing and formulation of paints is not complicated and quite possible for anyone to do in an emergency. White lead, or any lead-based paint, is almost impossible to purchase in the U.S., but is available in most other countries.

Lithopone is a combination of zinc sulfide and barium sulfate, and is never used with white lead pigments or with oils that use a lead dryer. This material is also available in most countries, and is used primarily on interiors because, when exposed to sunlight, it will darken.

Extenders are chemically stable, inert substances that do not react with the other components of the paint. These include aluminum silicate (China clay), calcium sulfate (gypsum), calcium carbonate (Paris white), magnesium silicate (asbestos), barium sulfate (barytes), and silica (silicious earth).

The *vehicles* that carry the pigments are drying oils such as linseed oil, tung oil, sunflower oil, and fish oil. The semi- and non-drying oils are cottonseed oil, corn oil, and soybean oil. The volatile oils and thinners are turpentine (pure gum), wood turpentine, benzol, toluol, coal tar naphtha, and alcohol. Turpentines are slow to evaporate, and they are also oxidizers. When used generously to cut the oils, they result in a flat paint or primer. Mineral spirits are derivatives of crude oil and completely evaporate as the paint dries. Coal tar naphtha, benzol, and toluol are derivatives of distilled coal tar.

Of the *dryers* used, the oil types such as lead oxide, manganese oxide, and lead-manganese oxide are dissolved in turpentine. The liquid dryers combine with the oil dryers but contain turpentine, benzene, or both. Japan dryers are the most common of the liquid dryers, and contain gum resins.

One gallon of red lead paint is made by mixing 2½ quarts of raw linseed oil and ½ pint of mineral spirits with 20 pounds of red lead. Coverage is about 500 square feet per gallon. Red lead is used as a primer over clean, wire-brushed steel.

One gallon of anticorrosive paint is made by mixing 3 quarts of grain alcohol with one pound of gum shellac, then mixing in ½ pint of pine tar oil, ½ pint of turpentine, one pound of dry metallic zinc, and 3 pounds of dry zinc oxide. Coverage is about 250 square feet per gallon. The above formula becomes an antifouling paint with the addition of ½ to ⅔ pound of copper flakes, precipitated and dry. Coverage is about 200 square feet per gallon.

Black topside paint can be made by mixing 4 pounds of lampblack in oil, then adding 3½ pints of raw linseed oil, ¾ pint mineral spirits, and one pint of benzene. Coverage is about 500 square feet per gallon, and this paint is always laid over anticorrosive paints or red lead paints.

Any of the drying oils may be substituted for linseed oil in the above formulas.

PAINT MANUFACTURERS

Ameron Protective Coatings Division
Ameron, Inc.
201 N. Berry St.
Brea, CA 92621
zinc-rich coatings, Dimetcote

Boat Life Manufacturing Company
205 Sweet Hollow Rd.
Old Bethpage, NY 11804
marine caulking compounds

BYCO Bywater Coatings
709 Engineers Rd.
Belle Chasse, LA 70037
general marine coatings

Devoe & Raynolds Company
Division of Grow Group, Inc.
4000 DuPont Circle
Louisville, KY 40207
complete marine coatings

Epifanes
Coastaltrade, Inc.
601 S. Andrews Ave.
Fort Lauderdale, FL 33301
marine coatings, primarily yacht finishes

Fiberglass-Evercoat Company
6600 Cornell Rd.
Cincinnati, OH 45242
epoxy and polyester resins

Hempel's Marine Paints
65 Broadway
New York, NY 10006
complete marine coatings

International Paint Co., Inc.
Morris and Elmwood Aves.
P.O. Box 386
Union, NJ 07083
complete marine and industrial coatings

Jotun-Baltimore Copper Paint
501 Key Hwy.
Baltimore, MD 21230
complete marine coatings and glassflake; yacht finishes sold under trade name Regatta Paints

Kenco
Division of Southern Coatings
P.O. Box 160
Sumter, SC 29150
cold galvanizing compounds

Koppers Chemicals and Coatings, Inc.
Koppers Building
Grant St.
Pittsburgh, PA 15219
complete marine and industrial coatings; yacht finishes sold under trade name Z-SPAR

Kristal Kraft, Inc.
Palmetto, FL 33561
resins, fillers

Pettit Paint Co., Inc.
36 Pine St.
P.O. Box 378
Rockaway, NJ 07866
complete marine coatings

Philadelphia Resins Corporation
20 Commerce Dr.
Montgomeryville, PA 18936
chocking materials

Porter Paint Company
13th and Walnut Sts.
Louisville, KY 40201
complete marine coatings

Proreco
PRC
5454 San Fernando Rd.
Glendale, CA 91203
deck coatings

Rust-Oleum Corporation
11 Hawthorne Pkwy.
Vernon Hills, IL 60061
protective primers and coatings

Salwico Glassflake, Inc.
5 Marineview Plaza
Hoboken, NJ 07030
Glassflake

Sherwin-Williams Company
101 Prospect Ave.
Cleveland, OH 44115
complete marine and industrial coatings

Sigma Coatings, Inc.
P.O. Box 826
Harvey, LA 70059
marine and industrial coatings

Texaco, Inc.
International Marine Sales Dept.
2000 Westchester Ave.
White Plains, NY 10650
Floatcoat, tank coatings

Transocean Marine Paint Association
Central Office
Mathenessiaan 300
Rotterdam, Holland
marine coatings

Travaco Laboratories, Inc.
345-T Eastern Ave.
Chelsea, MA 02150
epoxy coatings, potting compounds, Marine-Tex

U.S. Paint
Division of Grow Group, Inc.
831 South 21st St.
St. Louis, MO 63103
marine coatings sold under name of Awlgrip

Woolsey Marine Paints
183 Lorraine St.
Brooklyn, NY 11231
complete marine coatings and glassflake

7

$\triangledown$

LAUNCHING

The preparations for launching have been in the back of the builder's mind since the day the keel was laid. These preparations have not dominated his thinking, of course, because if he were to spend all his time worrying about how to launch the vessel, it would never be built. In the long sequence of problems that must be solved, launching is but one more. From time to time during construction, however, certain peculiarities of the vessel become apparent, peculiarities that will affect the launching procedures. These the builder files in his memory for future reference, and when the time comes, total recall is instantaneous.

The time-honored method of launching was to construct a building ways atop numerous pilings, these being driven into the ground parallel to the centerline of the vessel and 18 to 24 inches on either side of it. Each pair of piles was then capped with a crosstimber, and on these caps were set the keel blocks, which supported the keel during construction. The building ways were constructed at the same declivity as a second set of ways, which was offset about one-sixth the vessel's beam on either side of the centerline. These were the ground ways, upon which were constructed sliding ways to actually launch the vessel. Yards that build within the size limit covered by this book—less than 80 feet on deck—seldom employ this method any longer, since the expense of constructing individual building, ground, and launching ways is considerable, draining capital that may better be used in the purchase of equipment *and* in providing an alternate means of launching, such as a railway. Furthermore, when there is limited waterfront space available, using it for one or more permanent ways, which cannot be used for other purposes, would increase the overhead of a builder. Nevertheless, this type of permanent ways is still used in some of the larger yards. The vessels are then launched light and taken to a fitting-out berth where, as the name implies, all outfitting and joinerwork is completed. The keel of another vessel is laid down on the building ways almost immediately.

From the old method we learn that the width of any launching cradle must be at least one-third the beam of the vessel in order to prevent her from falling over. Making the cradle wider does no harm, especially if vessels of different sizes are to use the same cradle and launching facilities.

To attempt to cover every possible launching situation would create an endless task; therefore, the ones covered here are those that I have used at various times. These can serve as points of departure in adapting a system to your particular situation. It is often comforting to know that one need not be a pioneer, forced to wonder whether a proposed method is untried; someone else has probably been in a similar predicament, and arrived at a similar conclusion. A little common sense and a bit of elementary physics, along with some physical exertion, solve most problems. Of course, the simple solution of hiring a crane to lift the vessel from the building ways directly into the water or onto a cradle or lowboy is a quick and viable method if the cost is within the available budget, and the same holds true for a mobile boatlift. Please do not infer, then, that I am suggesting anyone should jump in the well just to get a drink of water, or even that he should use a bucket to haul water up from the well. I do think, however, that there is nothing wrong with a hand pitcher pump.

The five basic possibilities of a construction site are:

(1) the vessel was built at or near the water's edge;
(2) the vessel was built at a distance from the water's edge but need not go under power lines or across county, state, or federal roads;
(3) the vessel was built inland and must be transported to the launching site over roads (and, probably, under power lines);
(4) the vessel was built on rented space in a commercial yard;
(5) the vessel was built in an established boatyard that has all the necessary launching facilities.

The actual launching site determines the type of cradle to be used as well as the method employed for launching. Factors such as the depth of water, range of tide, type of ground (sand, shale, mud, rocks), and width of deep water (river, creek, bay, open roadstead) must all be considered. Will the vessel go directly to a mooring, or will she be taken to an outfitting dock or towed to another area? Will she take on ballast and have her spars stepped before or after launching? These and other questions must be answered before actual preparations begin.

Using the Pinky as an example to explore the first four situations, one can deduce the probable best course of action to follow in a given set of conditions. The fifth situation is covered by standard yard practice.

Vessel Built at or near Water's Edge

If there is a tide of 4 feet or more, it will be possible to launch the Pinky in a standard sled-type cradle in the upright position, keeping the keel in contact with the launching cradle. The builder can roll her out on the beach at low tide, remove the rollers, and simply wait for the next high tide. Without ballast, stores, water, and

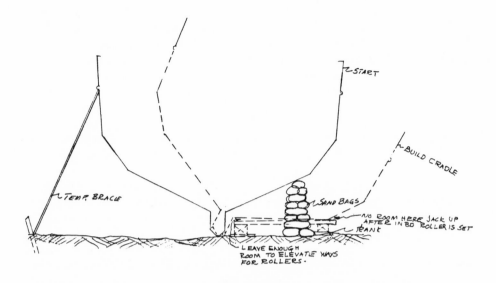

Figure 40. *A technique for laying the Pinky over onto a short cradle beneath her bilge. Slitting the sandbags tier by tier allows the vessel to settle in a controlled fashion.*

spars, she should float free. If the issue is in doubt, however, the builder may add a little ballast at the deck edge forward of amidships on one side, and then remove the after brace on the same side. This combination should permit her to be slued out, and she will float in even less water.

Should there be little or no tide, and should the distance to deep water be, say, 200 yards, a temporary ways must be laid and ballasted down. In this circumstance, an investment in four V-grooved wheels would permit the cradle to be fitted to angle tracks that could be leap-frogged out using two 20-foot sections. The workboat should be attached to the bowsprit as soon as possible after the vessel is clear of the shore, and loaded with several hundred pounds of ballast. Keeping the boat snugged up will keep the bow of the vessel down, thereby reducing her draft and aiding the builder in moving her as she gains buoyancy. When deep water is reached, the falls to the workboat are gradually eased until the vessel is afloat.

If the distance is short, say about 50 yards, it would be more economical to lay her on her side and use a short cradle under the bilge, simply sliding her out on ballasted planks with pipe rollers. A vessel may be laid over easily by using several tiers of sandbags to build up a cushion under one side to within 12 inches of her lower chine. The builder then removes the shores on that side and pushes her over to rest on the sandbags. (Too much of a drop will cause the keel to slide out.) When the sandbags are slit, the vessel will settle in a fully controlled fashion, tier by tier, until just the last couple of tiers remain, and there is room enough to slide the cradle in. With the cradle fitted and the pipe rollers inserted, the remaining bags are cut and the vessel is free to slide. Ballasted planks are laid ahead of her, and stakes are pushed into the bottom along the edge of the planks to guide the builder lest the vessel go astray. The best procedure is to lay the vessel's anchors well out and bend several warps together for a single lead to the windlass. This allows one to pull the vessel out over the pipe rollers until she is afloat. It is possible to do this working alone, but an extra hand or two is always welcome in *any* beach launching. The cradle must be lashed to the hull by running lines to the bulwarks near the forward and after ends of the cradle. These lines should be independent; otherwise, the cradle will slide out as the weight of the hull lessens. Again the workboat is used, this time to prevent the hull from coming up on an even keel before deep water is reached. The boat is oriented longitudinally, with part of its bottom resting on the vessel's railcap and part on the vessel's deck. It is filled with water as required to keep the vessel heeled, and once in deep water, it is emptied either by pulling the plug or with a siphon hose. The vessel must always be kept well heeled in shallow-water launchings; therefore, a good plan is to step and set up the spars at this time; one can then suspend the workboat from the throat halyards, thus lengthening the fulcrum.

Vessel Built at a Distance from Water's Edge, but Neither Power Lines nor Roads Complicate Route to Launching Site

Using a sled-type cradle with rollers moving along planks laid ahead of the vessel is the easiest and least expensive method for launching the Pinky. Pipe rollers are best to use, but good fence poles are satisfactory. No more than six rollers are needed, with three active rollers under the cradle, a fourth coming clear of the cradle behind, a fifth just engaging the cradle ahead, and a sixth laid in the cradle's path in case she rolls faster than expected and cannot be stopped. As the after roller comes free, it is brought forward and laid in position. A ditch in the path of the vessel can be bridged with a crib made of railroad ties or heavy timbers. Using this method, I have moved vessels larger than the Pinky almost a mile to the launching site.

A resting load on rollers requires a pull or a push equaling one-tenth its weight to start it moving, after which about one-fifth to one-tenth of the starting load is required to *keep* it moving. With sufficient help, the rollers can be kept ahead of the

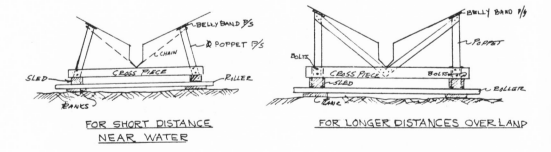

Figure 41. *Views from forward of a cradle for a short move and one for overland transport.*

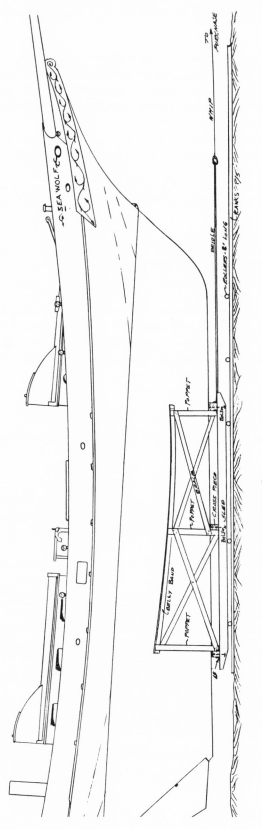

Figure 42. *Side view of an overland cradle.*

vessel, and she need never stop until the destination is reached. The rollers will not always roll straight, but all one need do is to tap the ends with a sledge hammer in the direction one wishes the vessel to go. To make an abrupt turn, bring the vessel to bear and balance on just one roller, then pull her around from both forward and aft, making sure that she does not drop off that roller or tilt off one edge. Insert another roller forward and one aft, and knock the center roller around until it is canted even more toward the desired direction of travel. Repeat this process until the vessel is on her new course, then square the rollers to the direction of travel, and proceed as before.

With large vessels, it is seldom possible to have full control of the vessel on a direct pull unless a large tractor or bulldozer is available that can be downshifted to a creeping mode. One must avoid jerking the vessel, since a violent jerk will either destroy the cradle or pull the vessel over. If the only power is manpower, the builder will need to plant the vessel's anchors well ahead and in tandem. A long rode is then led toward the vessel, and a twofold purchase is attached to the cradle. When the purchase is two-blocked, it is then overhauled and refastened to the rode for the next pull. A horse or mule may necessitate this same procedure, since these beasts have a tendency to hunch into the load and jerk the vessel from a standstill. Oldtimers have told me that a good yoke of oxen could drag a 50-foot vessel over smooth ground on a sled without rollers. Perhaps this is true, for pictures of such oxen indicate they were enormous beasts.

Vessel Built Inland

This scenario calls for the services of commercial haulers that specialize in moving boats or large loads. The builder, in most cases, will need to make a sled cradle that will carry the vessel during hauling and moving onto the lowboy. When the distance is of any magnitude, and especially in hilly or mountainous areas, this is the best solution. The trucker will arrange for the necessary permits and escorts for a wide load, and devise ways to reach the destination without having to move many power lines.

For a short haul the builder may elect to build his own trailer and assume the risks of moving. If he does, he may well begin by obtaining a set of tandem axles with good tires from a wrecking yard. A plate is attached to the axles longitudinally, and to the plate is welded a channel, angle, or other fixture that will accept and secure the keel. This rig is slid under the vessel a bit aft of the hull's longitudinal center of gravity, assuming she is to be transported bow first. The weight of the hull is borne partly by the channel and partly by the springs on either side via a fore-and-aft timber, from which supports extend upward to the bottom of the hull. A fifth wheel attachment is welded to the bow or forefoot; this wheel supports the bow but does not carry much hull weight. The alternative is to extend a tongue from the axles to provide an attachment for the fifth wheel fixture, in which case the bow is chained to the tongue.

Use of a farm wagon with front wheels that turn for changes of direction is also a possibility, the main precaution being to ensure the wheels are aligned. Otherwise, the whole wagon will sheer, weave, and eventually self-destruct while being towed. Only in rare instances would the vessel be transported to any place other than a facility that could launch her properly; however, if a beach, riverbed, or embankment has the proper slope, there is nothing to prevent using such a trailer wagon as a beach cart. On trailers using the fifth wheel, the builder would need to substitute a nose wheel for the actual launching, except on a steep grade into deep water.

Vessel Built on Rented Space in a Commercial Yard

In this instance, the approach depends on the equipment available. If the yard does not have heavy lift capabilities, it may be possible to build the vessel either directly over or adjacent to the far end of one of the launching tracks. In the first instance, the standard yard launching cradle can then be worked under the vessel from its end; in the second instance, the procedure would be to slide the vessel over and onto the cradle, or to remove the cradle from the tracks, work it under the hull, and move vessel and cradle back onto the tracks. Some yards have a concrete surface and can use carriages with swiveling wheels to get the vessel onto the tracks. If the vessel is not too large (and the Pinky is not), a mobile boatlift could be used, this being the most expeditious launching method when available.

MOVING THE VESSEL ONTO A CRADLE

The Pinky was constructed using three supports to the keel, which elevated it to a height that was convenient for welding and other work around the hull. Unless the vessel is to be lifted out of these supports by a crane or a mobile boatlift, it now becomes necessary to construct a cradle and move her onto it for launching, for transporting her, or both. This is accomplished in the following manner.

To begin, a stout vertical timber is clamped over the bulwarks on either side of the vessel abeam of the center support. The center support then becomes the pivoting point about which the vessel is brought to an even keel. The vertical timbers should be set with their bottom ends a bit forward so they will be correctly oriented when the bow drops and the keel is leveled. A further precaution is to secure the lower ends from side to side to prevent their skewing out should there be a settling or rocking of the hull. At or near the forward and after keel supports, another pair of timbers is clamped port and starboard (2 x 6s will suffice). Under the forebody, just aft of the forward supports, the builder lays up wooden blocks until they reach the keel and are snug and stable. A block of wood about 12 inches long is cut from a 6 x 6, V-notched to fit the keel, and placed atop the jack that will be used to lower the vessel to where the keel is level. (A jack that is rated at, say, five tons will support five tons but will not move it, so one should be careful to select a jack of sufficient reserve capacity to counteract any loading that might be incurred.) Another block, not as sturdy, will be needed aft. When using jacks for lifting, steel is never placed directly in contact with steel, since there is a tendency for one piece of metal to slide on another, especially when they are inclined. A block of wood, no matter how thin, is

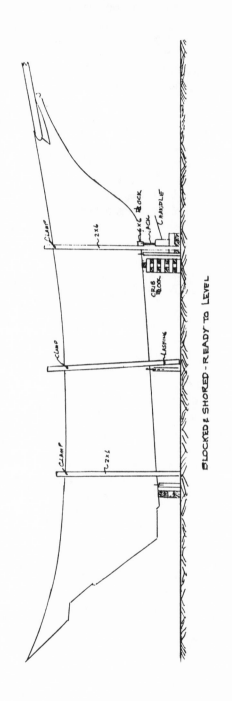

BLOCKED & SHORED - READY TO LEVEL

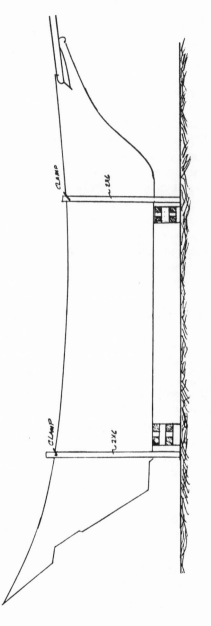

LEVELED - READY FOR CRADLE

Figure 43. *Two steps in the readying of the Pinky for a cradle, as described in the text.*

always used between the jack and the vessel. This also applies for a steel launching carriage. Wooden blocks must be placed on the carriage under the keel, since there is a tendency for the vessel to slide off if the incline is steep and the carriage must be stopped during launching.

The vessel is in no danger of capsizing as long as it remains in a near-vertical position, this being the purpose of the center supports. From this time onward, however, until the keel is level, one should approach the vessel only from the ends just in case it holds a grudge. The builder will place the jack (with the block on top) just forward of the forward support in order to take a strain and lift the vessel, so there is no weight on the forward support. When this is accomplished, the forward support is cut away, and any temporary bracing that is supporting the vessel forward of the center keel support is removed. The jack is then eased off until the keel rests on the forward support blocks. If the keel does not settle onto the blocks, it will be necessary to use a jack aft to pivot the keel down forward. If the keel does come to the blocks, the builder may rest assured that it will always do so. The next item of business is to take a strain on the jack once again, and, once the keel blocks are loose, adjust the forward braces upward until their bottom ends clear the ground by about the depth of the upper keel block. Remove the upper block and then release the pressure on the jack handsomely (smoothly) until the keel is again resting on the blocks. The after supports are next adjusted until they are firmly on the ground. The builder checks the midship supports to ascertain that they are secure and mutually aligned, and then repeats the whole procedure again, as many times as necessary to bring the keel to a level position. When the keel is leveled, it is blocked well aft of the center support, and the forward and after pairs of supports are firmly set into the ground. Once more, just enough of a strain is placed on the jack to ease the weight off the center support, which is then cut away. When it is clear of the vessel, the after keel support is cut away. Lifts need not be of great magnitude. Most often, an eighth of an inch or so is all that is needed.

The builder can now slip the cradle under the vessel or, if it has not yet been made, construct it in place. The bottom runners should be at least a 6 x 6, and preferably a 6 x 8, laid parallel to the keel and about a sixth of the vessel's beam out on the port and starboard sides. The cradle length need not exceed half the length of the vessel, with one-third the length on deck being the minimum. In the case of the Pinky, since the bowsprit and stern overhang constitute some extra weight and length, the runners would be 16 feet. These are elevated and blocked clear of the ground but kept far enough below the keel to leave room for sliding the crosspieces through. About 12 inches in from each end, a 6 x 6 about 6 feet long is placed across the tops of the runners. One or two additional crosspieces are placed between the two ends and squared to the centerline, and all the crosspieces are bolted to and through the runners with two ⅝- or ¾-inch carriage bolts, the heads of which are on the bottom side. To the undersides of the crosspieces are fitted diagonal braces of 2 x 4 material to keep the cradle from wracking. On both endpieces an angle iron clip is lag-bolted fore and aft and spiked to the crosspiece. This permits a bridle to be attached for pulling the vessel in either direction.

The builder now elevates the cradle until the crosspieces are firmly against the keel and are level athwartships, at which point he blocks up the runners on each end.

Next he erects braces to support the bilge, one brace at each crosspiece. If the vessel must be hauled over the ground for any distance or will be launched from a beach, a V-brace is desirable. If she is to be placed directly on a launching carriage, upright braces (poppets) are sufficient as long as there is a belly timber running the length of the poppets on either side, and the poppets are secured below the keel. The poppets are secured to the crosspieces with plates on each side, which are lag-bolted both to the crosspiece and to the poppet. Old fire hose is a good material to cushion and prevent damage to the bottom paint. If this is not available, inner tubes split and nailed where the poppets or belly timber are in contact with the hull work well. Barring this, the builder may make his own padding with rags overlaid with canvas. The poppets must be a snug fit with the hull.

When all is secure, the temporary posts and braces are removed, and the vessel is ready to be lowered onto the carriage or ground planks. The jack is now placed under either the forward or the after crosspiece, on the centerline and directly under the keel. A strain is taken, lifting the cradle enough to take the strain off the end blocks. If the vessel does not lift evenly in the athwartship direction, it should be set back on the blocks and the jack repositioned. The lowering takes place in several stages, working alternately fore and aft. If too great a drop is attempted in one step, the blocks will capsize at the opposite end. Never remove more blocking than is needed for the drop, and always lower away handsomely, for, if the lowering is sudden, there is the possibility of jarring the blocking loose. Prior to the last drop, planks are laid below each runner and the rollers are positioned and held from rolling with small wooden wedges. The last drop should be small—3 or 4 inches at most. The other drops should not exceed 8 inches. The vessel may now be moved to the launching site, as has been previously described.

FINAL PREPARATIONS

Upon arrival at the launching site and prior to launching, all through hulls should be closed, portlights dogged down, the steering gear lashed, and companionway doors and hatches closed. Mooring lines should be ranged and secured to the vessel with one end of each line over the rail.

Stepping of the spars can be done either just prior to launching or after the vessel is afloat. There are advantages to each method, but generally, unless there is a means to lift the spars independently of the vessel (such as with a crane or a convenient tree), it is best to wait until after launching. There is an exception, that being when the vessel is to be laid on her side for launching, as mentioned earlier. Then the sheers used during the construction can be set up, the spars eased into place, and the rigging temporarily set up using the peak halyards. (The throat halyards are reserved for the workboat.) With this exception, stepping the spars after launching is easier, saving the builder from having to lift the spars from the ground to the deck and then reposition the sheers for stepping.

Spars can be stepped when afloat by moving the vessel within reach of a crane, by lying alongside another vessel whose mast and rig will permit the required lift (another advantage of a gaff rig, incidentally), by lying alongside a bridge (if the bridge tender is friendly), or by using the sheers.

Some ballast will be needed soon after the vessel is afloat, while the builder is finishing the rigging and bending on the sails and prior to ballasting for the trial runs, as discussed in Chapter 5.

A final thought: one should never forget that the vessel can be moved, lifted, and slued around by prizing with wooden and metal poles, bars, and blocks of wood. A convenient size to use for prizing is 4 x 4 hardwood in a 4- to 8-foot length, such as the oak separating pieces that often come with the load of steel.

8

▽

ALTERNATIVE METHODS, ALTERATIONS, AND FORMULAS

ALTERNATIVE METHODS

A vessel can be built outdoors on a concrete floor or pad, in a shop that has a concrete floor and overhead cranes, in a hangar, in a narrow wooden building or shed so constructed that it is possible to use the building itself for squaring and leveling frames, in a barn with a dirt floor—you name it. In a large industrial area, such jobs as bending deck beams or pipe or the stem can be job-shopped out, an often economical alternative that saves the builder a great deal of time (labor) and the expense of specialized tools. Sometimes a new boatbuilding yard will be on the former premises of a company engaged in another type of fabrication, and the builder may be able to adapt many of those tools and fixtures for his purposes. The possibilities are really unlimited. If a vessel can be built under cover with all the modern facilities and conveniences available, one would be foolish to build it outdoors using the methods our ancestors were forced to employ. Knowing how things used to be done, however, makes it easier for one to accomplish the same things with better and more modern facilities.

Throughout Volumes 1 and 2 of *Steel Boatbuilding*, I have assumed that you, the builder, will choose a site, make the minimum necessary alterations, build a mold loft and the framing platen, erect a temporary shelter, and so on, and that, upon completion, you will restore the site to its original state and abandon it as far as boatbuilding is concerned. I have also assumed that the vessel will be completed on that site, that you own or will purchase the necessary tools, and that you will later sell those tools or incorporate them into your home workshop.

These assumptions most accurately apply to the backyard or one-time builder who works only parttime on his vessel and faces a project spanning years rather than months. There are exceptions to all assumptions, and these should be considered.

Moving the Vessel Before Completion

When the monthly or weekly rental cost of a suitable building site is significant, it behooves the builder to work as fast as possible to complete the hull. He may then move or launch the vessel for completion at another location or in the water. In such a case, all steelwork and sandblasting and some painting is done, including the first coats of finish color on the interior and exterior. Nonskid is not yet added to the decks, but the bottom will have at least two coats of antifouling if the vessel is to be launched. All other finishing and work is delayed until after the vessel is moved. An advantage to this method is that the vessel is lighter and easier to move than it would be after the joinerwork were completed. When a vessel is launched before completion, it is customary to fit and align the propeller shaft and the propeller even though the engine cannot yet be installed. This eliminates the need to haul again prior to the final bottom painting. Finish coating of the topsides can often be done from a raft or float.

During the framing and construction, the designed waterline will have been punched or scribed at each frame. Before the vessel is moved, however, the builder must establish level lines or marks within the hull so that it will not be necessary to relevel the hull after moving. When working on a vessel that is not level, the builder need only refer to the level line and compensate for the list by placing a wedge under one edge of his level near the end; another wedge under the opposite edge of the same end will compensate for trim. Thus, if the builder determines that the vessel is listing $\frac{1}{16}$ inch per foot to starboard in the morning, the level is adjusted accordingly, and all subsequent thwartships leveling steps will compensate automatically for the listing. The level is rechecked as each new major piece of joinerwork is added. Once a plane (surface) is established, one may square from and off it. Building the interior afloat is in practice often less time-consuming than building it ashore, especially if all materials must be carried up ladders ashore. After all, one can walk on and off a docked vessel over a proper gangway.

Rented Space in Commercial Yards

There are itinerant builders who, for many reasons, do not wish to own their own facilities, preferring to rent space in an established yard where they can construct one or more vessels to their own account. The builder then employs yard personnel as needed. Basic materials are purchased through the yard on a cost-plus basis, and only the specialized materials are purchased from outside sources; the same understanding applies to the employment of outside labor. The leasing of facilities is normally considered by yards when they do not build that type of vessel; when the vessel is too small in relation to the yard's overhead to make the yard competitive in bidding; or when the yard is too large to be bothered with so small a contract (a small yacht, for example, when the yard normally builds 200-foot commercial vessels). The itinerant builder takes up some of the slack in the usage of labor and facilities that always exists in boat- and shipyards, and the yard employees may well learn new methods and techniques. Then, too, it is added income with no risk of capital.

The builder gains a mold loft, power tools, competitive purchasing power, and a trained labor force. He has no nonproductive administrators to pay directly, and no insurance, withholding, or other bookkeeping chores to worry about. When his work is completed, he can pack up his hand tools and move on. He is never in competition with the yard he rents from. A competent builder finds this a profitable solution; the incompetent builder may find it expensive, and in fact, whether he is building for others or on his own account, he can wind up bankrupt. Needless to say, it is not a solution for the first-time or one-time builder who may need advice, time, or extra workmen to make up for his lack of experience.

Repair yards, too, will sometimes rent space to the itinerant builder and supply him with the necessary large power tools and access to shops. Again, the builder is expected to utilize no outside labor or vendors. A repair yard, however, is more of a feast and famine operation than a building yard. Often, because of the prevailing work load, no employees are available to the builder; they may even be pulled off his work without warning to undertake a rush repair job. Meanwhile, his rental expenses continue. It is best, in a repair yard, if the builder can work alone for the most part, using his own portable power tools. This keeps costs down if the rental is not excessive.

Building with Level Keel

Due to height or other space limitations, or even the sheer size of the vessel, it is sometimes necessary or desirable to build a vessel with its keel level and its frames set at an angle to the world. In other words, the frames are level athwartships but are not plumb. It then becomes necessary to adjust one edge of the level with a wedge to correct for the declivity, or to employ a declivity board, which corrects for the angle between the slope of the keel as laid and the designed waterline when afloat. On the mold loft, the measurements for the ends must be corrected by extending the bottom of the keel so that the perpendiculars are square to the keel line when a plumb bob is dropped, rather than to the designed waterline, as is normal. The measurements along the keel are laid off directly from the mold loft, and all frames bear the same relation to each other as when the designed waterline is level. The designed waterline and all other waterlines are sloped, and they must be established in order to determine the correct frame heights, which are further checked with separate offsets measured square to the keel from the mold loft. The same procedure applies when a vessel is built on a building ways. It entails a bit more labor in the setting up of the frames, but poses no subsequent problems (apart from visual disorientation) as long as one remembers to use the modified level or declivity board. Although there are alternative methods, the painted waterline and the boottop line are usually centerpunched into the outside of the shell plating at each frame as the plates are applied, so that these lines may be struck after sandblasting.

Locating Waterlines

When a vessel is built with the waterlines horizontal, they are marked by filling a garden hose with water and fitting to each end a sighting glass or tube with a mark on it. One person holds one of the sighting marks of the water-filled hose at an established waterline mark, and a second person moves to any other portion of the hull and adjusts his sight mark vertically until the water in the hose intersects it. This position, which is scribed on the hull with a pencil or soapstone, is level with the first mark. The sequence is repeated until a sufficient number of marks have been established, at which time a thin batten may be used to connect the marks with a continuous line. This is a very quick method for marking a waterline, and one that can ignore irregularities in the terrain. In the extreme ends, it is often necessary to attach a short level to the sighting mark in order to transfer a mark onto the hull. Except in rare circumstances the waterline must be scribed with a rather narrow batten, since a wide one will not accommodate the compound curvature in either the fore-and-aft or athwartships direction.

Another method used on outdoor sites or in large shops employs a transit. This method, too, will ignore the terrain. If the transit's sighting scope can be elevated to the required height, marks can be made directly; one person with a soapstone merely moves it up or down as directed by the person sighting through the scope. If the scope cannot be elevated to the required height, a staff—like the staff used in surveying—will be necessary. The desired mark will be scribed at a measured distance above or below the sight mark. In another method of marking waterlines, the builder runs a long board athwartships at either end of the vessel, the top edge of both boards being level and at the same height as the waterline. Over the boards is stretched a string, which is gently eased in and against the hull. A mark is made where the string touches the hull. The string is then moved toward the centerline at one end and away from it at the other, and eased against the hull again, giving another point, and so on. By sequentially moving one end of the string out and the other end in, the whole forebody or afterbody is eventually marked. When one side is complete, the other side is done. The string cannot be wrapped around the hull. This method can be used whether or not the waterline is level with the earth; it works well on small vessels, but on very large ones the length of the boards becomes too unwieldy. In that case, pairs of posts with a piano wire strung between them can be substituted, the posts being erected well out from the centerline of the hull. The wire is tightened with a small turnbuckle, and strings can then be stretched or tightened over the wire. The danger is that any undue weight on the string not only pulls the wire out of position but can also cause it to sag. To reduce the distances the wire must span, one or more posts can be erected at intervals along the length of the hull and marked with the true waterline height. A wire is stretched across these marks on the posts just as was done with the end posts; the string may then be moved forward and aft on the longitudinal wires or in and out on the transverse end wires, as the case may be, to locate the required waterline marks on the hull.

Still another alternative is to run two strings in parallel 12 to 18 inches apart across the transverse end boards or wires. With one person sighting across the strings, a second one marks the hull wherever he is told. This demands a good eye and is quite tiresome to the person sighting the line. This method, however, can be used with complete satisfaction even if the waterlines are not level with the world.

ALTERATIONS

Lengthening

Once a vessel is built, she may remain as built for the rest of her life. Sometimes, however, alterations are made either to modernize a vessel or alter her for another occupation. It is possible, for example, to lengthen a hull by cutting it in half and splicing in a new piece. This may be done to fill a need for increased capacity, to correct a faulty design, or to alter the hull's powering characteristics. In a two-masted schooner, the piece added usually has a length equal to the distance between the foremast and the mainmast—in other words, 20 to 30 percent of the vessel's waterline length—and a third mast is added. The existing rigging, spars, and sails can then be used "as-is, where-is." Vessels having a straight keel parallel to the designed waterline lend themselves to stretching, the only major adjustment being the refairing of the sheerline in profile. On the other hand, vessels having drag to their keels will acquire an undesirable keel knuckle if they are pulled apart in such a manner that the waterline of each end remains in the same plane. Either the after section must be lowered to accommodate the extended midbody, or the forebody must be raised. Once this is accomplished, only a new sheerline need be faired in, but the change in trim will be quite drastic and will require the striking of a new waterline. The services of a naval architect should be employed for such an alteration, but seldom are.

When cutting a vessel in two for lengthening, one works with the maximum section. If the maximum section falls directly on a frame or bulkhead, the builder must decide whether the transition will be in the forebody or the afterbody. In theory, what is being added is a parallel and constant section, but in reality, this section must fair into the existing hull. Therefore, it will be either a continuation of the afterbody faired into the forebody, or a continuation of the forebody faired into the afterbody. When lengthening, it is generally easier to fair into the forebody. Either way, when the vessel is lengthened in this manner, the prismatic coefficient rises very quickly.

Occasionally there is reason to lengthen a vessel on the mold loft floor by adding one or two frames, making a proportional adjustment to mast locations, and enlarging the rig to suit. The procedure is no different than if the vessel had already been built. For example, in the case of the Pinky, one would lay off the forebody to frame number 7, add a space, lay out frame number 7 again, and then lay out the afterbody, fairing in the two halves through the new section. Because of the strong sheer of the Pinky, refairing of the deck must extend forward to approximately frame number 4 and aft to frame number 9 in profile. When two halves are pulled apart on the mold loft floor, they are pulled apart on the designed waterline. The builder has three options in that case: (1) He can use the forward and after ends of the straight portion of the keel as references, snapping a straight line between them with a chalkline and thereby changing only the amount of deadwood throughout the hull length. The draft does not increase and the center of lateral resistance is altered only slightly, but the center of pressure when sailing moves forward, since the keel has less drag. (2) He can continue the slope of the keel as established in the forebody, and by carrying it all the way aft, increase the vessel's draft. This increases the amount of deadwood, moves the center of lateral resistance aft, and moves the center of pressure aft when under sail. (3) He can hold the slope of the afterbody, extending it as a straight line forward. The draft remains the same, the amount of deadwood forward decreases, and the center of lateral resistance and the center of pressure both move aft. A vessel lengthened during original construction will always have a sweeter sheer than if it had been lengthened as a later alteration.

Another method of lengthening other than adding one or more frames to the midbody is to increase the frame spacing. This can only be done on the mold loft floor during original construction. The depth and width of the hull are not changed. When the frame spacing is increased in this fashion, *all* longitudinal measurements must be multiplied by the ratio of the new frame spacing to the old. Strongly raked ends such as those of the Pinky become even sharper, and the prismatic coefficient is reduced. Indeed, *anything* that is raked gains a more pronounced rake, while anything that is plumb is just spaced farther apart, so this type of alteration is usually limited to a vessel with very blunt ends. In sailing vessels the masts will be further separated, necessitating a complete modification of the sail plan. It is not customary to increase the height of the rig, but all else changes.

Adding a frame or section to a vessel increases her stability dramatically. Proportional stretching also increases stability, but to a much lesser extent. The amount of lengthening on the mold loft floor is generally limited to 10 percent or less of the deck length. Otherwise, it is better to choose a new design.

In sailing vessels, lengthening the hull not only affects sailing qualities but increases speed through the water. When the prismatic coefficient becomes higher, the vessel is required to sail faster in proportion to her increased length than she did before with her old length. With a lower prismatic coefficient, her optimum speed is lower. A drastic change in length almost always requires a change in rig. In a motor vessel, however, an increase in midship length requires no other alteration as long as her propulsion plant will handle the increase in tonnage. Unless there is a drastic increase in the prismatic coefficient, the motor vessel will usually have more speed with the same horsepower after she is lengthened. There is usually power to spare in these days of overpowered vessels, so the increased length is usually given over to an additional hold. Carried to the extreme, the increase in length would result in a self-propelled barge—a motor vessel 60 feet long, for example, becomes a barge when lengthened to 100 feet. Her depth and breadth do not change, but the constant sectional area in her middle raises her prismatic coefficient to that of a barge. When a vessel is lengthened with depth and breadth staying the same, the propulsion efficiency always increases. Done properly, lengthening will often take a vessel that is a dog as far as her propulsion characteristics are concerned and change her into an antelope.

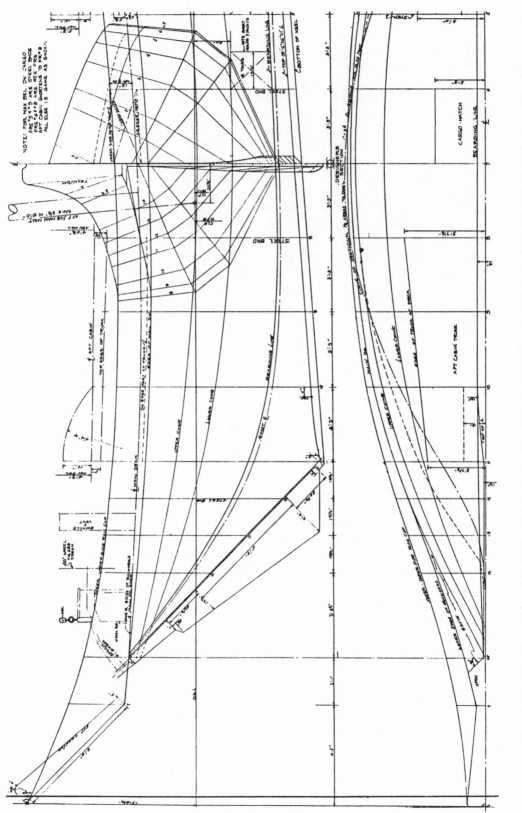

Figure 44. *The Lines Plan of the Pinky, Colvin Design No. 169, the building of which is described in the two volumes of Steel Boatbuilding. The accompanying text outlines how this vessel might be lengthened on the mold loft floor.*

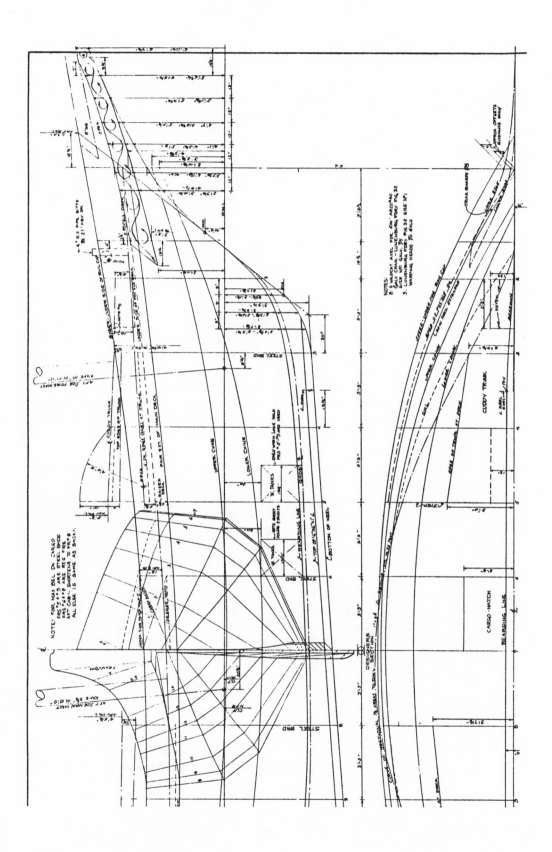

Shortening

Vessels are seldom shortened except to meet certain legal requirements such as the Limit Laws on fishing vessels, which permit the transfer of licenses from one vessel to another of the same length. When an existing vessel is shortened, it can enhance her appearance or make her downright ugly. The piece taken out is normally aft rather than forward, since a good entrance is always required. In vessels that have all of their machinery aft, however, shortening is usually done at the bow by cutting off all the rake to the stem above the designed waterline and either leaving the deck blunt or snubbing in the plates. In extreme cases, a tumblehome is incorporated in the stem profile, since in some instances deck length, and not overall length, is the criterion. A reverse stern and tumblehome bow can often take as much as 10 to 12 percent off the measured length on deck.

A New Bow

After a severe collision, a new bow might have to be added. Rather than restore the original shape, it is often desirable to design a completely new bow that lessens the angle of entrance and perhaps enhances the vessel's appearance. The procedure, in everything except a massive extension, is to set up the new stem at its proper location and cut away the forebody to a point where there is no damage. This must be between two frames. Wooden battens are fastened to the existing shell plating and pulled in to the new stem, at which time, via turnbuckles attached to the battens on the port and starboard sides, a new shape may be sprung that is pleasing to the eye of the builder.

This technique is borrowed directly from sawn frame wooden shipbuilding, where the mold frames were often six to ten frames apart. The builder dubbed off the frames a little here and there to get the shape he desired. While this may have been done by rack of eye, it certainly turned out some nice-looking vessels. In the case of steel, once the shape is decided upon, the builder makes a wooden pattern frame about halfway between the old portion of the hull and the new stem. This is faired in the usual manner and then replaced in the hull at its proper location. The battens are attached to this pattern frame, and patterns are made. All additional frames are then picked up as required directly from the hull. When the builder is satisfied that all are lying where they belong and are fair, they are made in steel and placed in their proper positions. Longitudinals are cut after the frames are up, the new longitudinals being always a fair extension of the existing ones. After the longitudinals and other parts are in place, the shell plating commences just as in new construction, with the same considerations.

Raising and Lowering the Deck

When a vessel changes from one trade to another, there are times when it is necessary to raise the deck in order to increase its capacity. The sheer is seldom altered; instead the whole deck structure is raised. The end result is hardly noticeable in slab-sided vessels except in the bow and stern, where knuckles must be added to accommodate the increase. In vessels with flare to their sides a continuous knuckle is formed from one end to the other, which ends up looking like a high chine nearly parallel to the deck.

Preparatory to raising a deck, it is customary to make the cut at least 12 inches below, but not necessarily parallel to, the existing deck edge. If, after the deck is raised, the result is just plain atrocious (and it can be), the portion between the knuckle and the old deck location is removed, and the hull above the cut is refaired into the new deck. In doing this, the services of a naval architect are usually employed.

In the best and least expensive procedure, the builder decides what the increase in depth is to be, then lays a long, solid template on the hull, arranged so that the plates to be added need have only one edge cut. This is easier in transom-sterned vessels than in cruiser- or fantail-sterned vessels, since in the latter types, there is a section of the hull over which one must twist the added plate and drastically increase its width in order to achieve the desired rise. The same holds true in a vessel with a strongly raked stem and great flare, as well as in a vessel with a stem having a soft nosing (the round at the deck and bulwarks molded in a conically shaped piece of steel).

Raising the deck requires numerous screw jacks and a great deal of cribbing. The jacks are set up with a slight strain. A level line is established athwartships in at least four places, and usually on every frame. The vessel is cut through at the desired mark, frames and all. The jacks are then opened up and the deck raised to the desired height. Of course, all wiring, piping, and other systems must be cut, as well as the bulkheads. In rebuilding the framework, it is customary to recut the frames after the separation and make the new frame joints above and below the actual welded seams. Again, if longitudinals are to be added, they are put in after the frame is in and before plating commences, and are run fair.

A more common alteration in older vessels—sometimes useful when the accommodations are aft or when repowering is desired—is to raise the quarterdeck—or in some instances, to add a quarterdeck by cutting the main deck and moving everything on it up to the level of the bulwarks. Generally, to be economical, the height of the quarterdeck is limited to the height of the bulwarks. If there is tumblehome in the bulwarks, cutting the deck after scribing it to the same half-breadth as the bulwarks will suffice. Both the deck plating and the deck beams are cut through, and the deck is elevated to the new position to be rewelded to the bulwarks, thus forming the quarterdeck. The framing and knees needed to join the pieces in this type of alteration are made as extensions from the lower frames, and the remaining, perimeter portion of the old deck is ground smooth and retained as a structural member.

If, on the other hand, there is flare to the bulwarks aft, it is best to make the cut so that the remaining, unplated portion of the new deck is at least 6 to 12 inches wide, in order that the welding of the new deck seam and the waterway plate can be done expeditiously. Some builders, in attempting to save every little bit of steel, end up with impossible welds and welds that will lead to maintenance problems later.

When the deck is raised, the center of gravity of the hull is raised as well.

Occasionally, a stability problem will require that a deck be lowered. This is especially true when superstructures are added to the original design. This alteration is more difficult than raising a deck unless the vessel is slab-sided. When it must be done at all, the vessel usually has other inherent problems as well, and the net result of the alteration is an adequate rather than a drastic improvement.

Other Alterations

One of the most common alterations made to a vessel is the changing of a deck arrangement or the modernizing of the superstructure. Unless excessive weights are involved with the new addition—for example, a large extension or an increase in height—the builder should encounter no problems. When a commercial vessel's deck arrangement is changed, the usual reason is new equipment or a change in occupation. Sometimes a commercial vessel is altered to become a yacht, or a yacht is altered to become a commercial vessel. The former is usually easier, because a commercial vessel probably already has sufficient depth and capacity, so additional weights need not cause undue worry. It must be remembered that weight above the waterline always detracts from stability, while weight below increases stability.

If a yacht *does* have sufficient volume for cargo and the buoyancy to carry additional weight, and if she has the right hull shape, the necessary deck alterations to make her a commercial vessel are usually quite easy to accomplish. Yachts traditionally have very long or large deckhouses, so the deckhouse will simply be cut at a convenient location, the unwanted portion removed, and the area decked over completely. A great deal of the interior furnishings must be removed, and structural bulkheads must be installed to accommodate the new hold. One must guard against major shifts in interior weight, such as moving the engine from amidships to the stern or the fuel tanks or water tanks into the extremities of the vessel.

Both power and sailing vessels are occasionally modified for the charter trade, which usually entails a major revision of the interior arrangement. Owner and builder need have no fear of attempting this in a steel vessel ashore or afloat, since, with rare exceptions, the structural integrity of a steel hull is independent of the interior joinerwork.

When one is considering *any* commercial use of a vessel that was built as a yacht, he would do well to consult with an admiralty lawyer. Suffice it to say that, under U.S. law, if a vessel not built in the United States is to fly the U.S. flag, it cannot be used in the charter business (except for bareboat charter), for coastwise freight, or for any other commercial purpose. There is one exception: a yacht may carry the personal cargo of her owner, who may buy, sell, trade, or do anything else he wishes with his own possessions; the owner cannot, however, carry freight or passengers for hire.

On occasion, a vessel built to travel rivers, bays, sounds, and other tributaries is altered to make it suitable for ocean passages. In such a case it is usual to modify the entire forward third of the vessel, and sometimes even more. The old section is removed and discarded, and a whole new forebody is added. Only the anchors, anchor chain, windlass, and items above the deck that may fit the new forebody are salvaged. This same modification is occasionally carried out to improve the propulsion characteristics of a hull, sometimes with significant success.

Once in awhile a stern is lengthened. The portion affected is normally at or above the waterline, and the goal is better propulsive efficiency or the correction of a trim deficiency. Some vessels—particularly some of those with fantail or cruiser sterns—squat so badly under power that freeboard diminishes dangerously or disappears altogether. By modifying the stern, squatting can be reduced or eliminated.

Power vessels with their propulsion machinery and accommodations entirely in the stern, such as small tankers, do find it economical to scrap the forebody of the vessel when it deteriorates to the point of requiring extensive repairs to the shell and framing. In reality, as much as two-thirds and sometimes three-fourths of an entire vessel may be scrapped due to deterioration caused by cargoes such as oil and caustic soda. At this time, the vessel may be rebuilt as originally designed (which is rare), or modified (which is more common). The modification possibly and probably will include a lengthening, since all that is being built forward of the after end is a compartment requiring very little piping, wiring, or anything else. The propulsion machinery, steering gear, auxiliaries, fuel tanks, water tanks, lifeboats, accommodations, navigation area, and so on are all contained in the end that is saved, which constitutes about 90 percent of the total cost of such a vessel.

In summary, a multitude of alterations is structurally possible, but an alteration is not always economically feasible. A new vessel would often be cheaper.

FORMULAS

A builder is forever being confronted with the need to convert information supplied by manufacturers, since none of them seems to use measurements that are compatible with existing yard practice or equipment. Table 11 presents an assortment of paired measurements, with the appropriate factors to use when converting one to another. A few formulas are also included, these being useful but not so frequently used that one must burden his mind by remembering all of them.

Table 12 merely gives a number of solutions for two of the formulas in Table 11, the object being to save you some calculations. For example, if the water intake to the engine is, say, 7 feet below the waterline, the pressure through the pipe, assuming it is horizontal, is 3.111 pounds per square inch (psi). This head may be sufficient for adding a bypass to the saltwater cooling system. Then, in the event of a water pump failure, cooling water may still reach the engine, from there to be pumped back overboard by a second pump. The table is also convenient for determining the amount of pressure required on a patch to cover a hole at a given depth. This is particularly useful when making a temporary patch on a vessel to move her to a docking facility.

Table 13 is included for those rare instances when a builder needs to solve a problem but lacks a formula.

TABLE 11

Conversion Factors and Formulas

atmospheres	$\times$	33.90	= feet of water
atmospheres	$\times$	29.92	= inches of mercury (Hg)
atmospheres	$\times$	14.70	= pounds per square inch (psi)
atmospheres	$\times$	2120.	= pounds per square foot
British thermal units (BTUs)	$\times$	778.30	= foot-pounds
BTUs per hour	$\times$	0.2931	= watts
BTUs per minute	$\times$	0.02356	= horsepower
degrees centigrade	$\times$	1.8 + 32	= degrees Fahrenheit
cubic feet	$\times$	7.48052	= U.S. gallons
cubic feet	$\times$	62.5	= pounds fresh water
cubic feet	$\times$	64.0	= pounds salt water
degrees Fahrenheit	$\times$	0.5555 − 32	= degrees centigrade
feet of water	$\times$	0.0295	= atmospheres
feet of water	$\times$	0.8826	= inches of mercury (Hg)
feet of water	$\times$	0.4335	= psi
U.S. gallons	$\times$	0.1337	= cubic feet
U.S. gallons	$\times$	0.8327	= imperial gallons
imperial gallons	$\times$	1.2009	= U.S. gallons
U.S. gallons water	$\times$	8.3453	= pounds of fresh water
U.S. gallons water	$\times$	8.6	= pounds of salt water
horsepower	$\times$	33,000.	= foot-pounds per minute
horsepower	$\times$	550.	= foot-pounds per second
horsepower	$\times$	745.70	= watts
inches of mercury (Hg)	$\times$	0.03342	= atmospheres
inches of mercury (Hg)	$\times$	1.133	= feet of water
inches of mercury (Hg)	$\times$	0.4912	= psi
liters	$\times$	0.2646	= U.S. gallons
pounds	$\times$	7000.	= grains
pounds	$\times$	453.5924	= grams
psi	$\times$	0.06804	= atmospheres
psi	$\times$	2.037	= feet of water

One ton of fresh water = 35.84 cu.ft. = 268 U.S. gallons.

One ton of salt water = 35.0 cu.ft. = 262 U.S. gallons.

Circumference of a circle = πD or 2πr, where D = diameter and r = radius.

Area of a circle = π D^2 ÷ 4, or π r^2, or D^2 $\times$ 0.7854, or C^2 $\times$ 0.07958, where C = circumference.

Length of any arc of a circle = $\dfrac{8c - C}{3}$ (Figure 45-A).

Area of an ellipse = a $\times$ b $\times$ 0.7854 (Figure 45-B).

Volume of four pipes crossing = π r^2 (L+L' − ⅔r) (Figure 45-C).

Pressure per square foot in fresh water = 62.5 $\times$ depth

Pressure per square inch in fresh water = 0.433 $\times$ depth (approx.)

Pressure per square foot in salt water = 64 $\times$ depth

Pressure per square inch in salt water = 0.444 $\times$ depth (approx.)

Absolute pressure is found by multiplying the area in square feet $\times$ 2120, or the area in square inches $\times$ 14.70, to correct for atmospheric pressure.

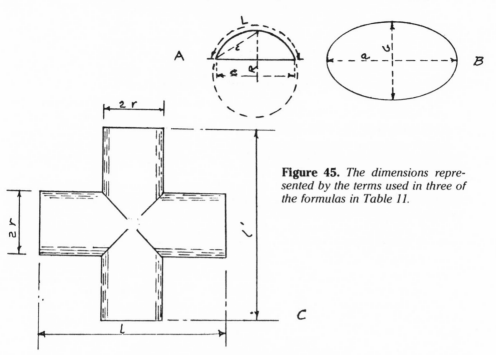

Figure 45. *The dimensions represented by the terms used in three of the formulas in Table 11.*

TABLE 12		
Pressure of Salt Water at Different Heads		
Head (feet)	**Pressure (lbs./sq. ft.)**	**Pressure (psi)**
1.0	64.0	0.4444
1.25	80.0	0.5555
1.5	96.0	0.6666
1.75	112.0	0.7777
2.0	128.0	0.8888
3.0	192.0	1.3333
4.0	256.0	1.7777
5.0	320.0	2.2221
6.0	384.0	2.6666
7.0	448.0	3.1110
8.0	512.0	3.5553
9.0	576.0	3.9998
10	640.0	4.4442
20	1280.0	8.8885
30	1920.0	13.3328
40	2560.0	17.7770
50	3200.0	22.2213
60	3840.0	26.6656
70	4480.0	31.1098
80	5119.8	35.5541
90	5759.8	39.9984

TABLE 13

Miscellaneous Formulas

Permissible stress in the upper deck $=\sqrt[3]{\text{LBP}}$, where LBP $=$ length between perpendiculars.

Period of roll (T) $= 0.42$ B $/ \sqrt{\text{GM}}$, where B $=$ beam and GM $=$ the distance from the center of gravity to the metacenter.

Period of wave (T) $= 0.442 \sqrt{\text{L}}$, where L $=$ wavelength.

Approximate bending moment $=$ displacement $\times$ LBP/30, where LBP $=$ length between perpendiculars.

TABLE 14

Conversion Factors in Measurement

fraction of an inch	decimal parts of an inch	decimal parts of a foot	inches	decimal parts of a foot
1/16	0.0625	0.005	1	0.08
1/8	0.125	0.01	2	0.17
3/16	0.188	0.016	3	0.25
1/4	0.25	0.02	4	0.33
5/16	0.3125	0.026	5	0.42
3/8	0.375	0.03	6	0.50
7/16	0.4375	0.037	7	0.58
1/2	0.50	0.04	8	0.67
9/16	0.5625	0.047	9	0.75
5/8	0.625	0.05	10	0.83
11/16	0.6875	0.057	11	0.92
3/4	0.75	0.06	12	1.00
13/16	0.8125	0.068		
7/8	0.875	0.073		
15/16	0.9375	0.078		

Decimals are rounded off here as they are commonly used by boatbuilders. For example, 10⅝ inches converts to 0.83 foot + 0.05 foot = 0.88 foot; actually, it is 0.8854 foot. In normal practice, with the many numbers used in a calculation, the average is quite close to the mythical absolute. It is also common to add 0.01 foot to an occasional second-place decimal to compensate for the frequent dropping of a third-place decimal that is greater than .005 foot. Decimals are used in machine shop and plating designations and should be memorized by a builder to save time in working with catalogs, pipe diameters, etc.

TABLE 15

Mathematical Symbols

For some reason, much of the information that is sent to a builder is loaded with mathematical symbols. These probably mean something to those who have had the new math, but may not hold much meaning to the many who learned the old way. The most common ones follow.

Symbol	Means	Example
=	equal	A = B
<	less than	A < B
>	greater than	A > B
< >	is not equal to	A < > B
< =	is less than or equal to	A < = B
> =	is greater than or equal to	A > = B

Boatbuilding involves a host of rules of thumb, and it often seems that a builder needs at least 10 thumbs. All builders will find it convenient to be able to convert mentally from fractions to decimals, from decimals of an inch to decimals of a foot, and from inches to decimals of a foot. Table 14 shows the approximations used in boat- and shipbuilding yards. There is no practical reason to carry calculations to six or seven decimal places when all measurements are made to the nearest eighth, or at best, to the nearest sixteenth of an inch. Most of these conversion tables, of course, could be eliminated by using the metric system, but until such time as everyone adopts the metric system, it is better to think in the old way, for it is easier to convert *to* metric than to convert *from* metric.

9

$\triangledown$

BOATBUILDING
AS A BUSINESS

Boatbuilding can be one of the most satisfying and creative occupations, but it is also one of the most challenging. The builder must use his hands and brain to the fullest in order to survive. There are those who own boatyards, but who are not builders. They are businessmen, adept at managing personnel. To them the product is immaterial; it is just a question of profit and loss—today, boats; tomorrow, aircraft, houses, furniture, or whatever seems the most profitable. Quantity is valued more than quality. In farming, this type of person is known as a "rump" farmer. So, one may ask, how does one define a real boatbuilder?

The rare individual who possesses the capability, determination, and skills necessary to construct a wooden or metal boat with his own hands and by himself is recognized the world over by the marine industry and governments as a master carpenter or master builder. Indeed, when he has completed a vessel, he will issue a Master Carpenter's Certificate or Master Builder's Certificate attesting to the dimensions and setting forth other pertinent facts that describe and otherwise identify the vessel. The certificate eliminates any confusion as to what is rightfully his creation. When he begins construction of a vessel, the builder need not be a master of all the trades that go into building it; by the time the vessel has been completed, however, he will have become competent if not proficient in all these trades.

Boatbuilding is a creative art form, but not everyone is creative. The so-called amateur boatbuilder is a rare, tenacious individual who realizes that, given time, he can assimilate the necessary knowledge and craftsmanship and can construct a vessel that will not only be structurally sound, but will float and function as designed. There are thousands of persons employed in the boat and shipbuilding industry who have spent or will spend their lives constructing vessels, yet few ever have the drive or desire necessary to accept total responsibility for a vessel as a master builder must accept it. Indeed, in many boat and shipyards the yard itself becomes the master builder, since a vessel is assembled from the collective efforts of individual departments, with a corporate management taking the credit. When a deficiency is discovered in the completed vessel, it is "buck-passing" time in the building yard. The individual entrepreneur, on the other hand, is the master builder, and he is totally responsible for the vessel just as the captain of a ship is totally responsible for his command. While he can delegate authority, he cannot delegate responsibility.

A full-time boatbuilder and yard owner has usually become one through a process of evolution. Having built one vessel, usually for himself, he develops an insatiable desire to build another, and then another. The next thing he knows, he is in the boatbuilding business. To be quite truthful, this is one of the few businesses in which one must have a vast working knowledge, be competent with his hands, and at the same time possess business acumen. One is in business to make a profit, but in the boatbuilding business profits are sometimes elusive, at least at first. A builder of 100 good boats is just a boatbuilder, but if he should build one bad boat, he becomes known as a bad or "lousy" boatbuilder. Like the farmer standing up to his knees in "manure," the boatbuilder has his feet planted on the ground, and there is no point in telling him the difference between that and a bed of roses. A house may hide a deficiency in construction for years, but any major deficiency in the structure of a vessel shows up almost immediately upon launching.

Unless an individual boatbuilder has a wealthy wife or partner, a large inheritance, or a "sugar daddy," his equipment must accumulate gradually. If he is to remain solvent, the purchase of any new piece of equipment is always based on its ability to pay for itself during the first job. If he keeps on purchasing tools, always wanting the ultimate tool regardless of cost, he can become tool poor, and any profit he may make goes right down the drain.

One of the first problems encountered by anyone wishing to enter the boatbuilding business is how to disburse the very limited capital that is available. Overall insurance for the yard and its equipment is out of the question because of the huge expense involved. The builder may elect to insure each vessel under construction as an element of its contract price, or he may gamble that his building and maintenance procedures will protect him from the danger of fire, accidents, and weather. This gamble is often necessary in order to be competitive, but one major catastrophe can end such an enterprise. In the ideal business, one ends up with 10 percent more than he has expended.

The establishment of a good credit rating is one of the major requirements of any business. The easiest way to do this is always to pay bills on time even if this demands personal sacrifices, because when your word becomes your bond, your credit is always good, and there is no delay in shipment or question of payment. One of the easiest ways to destroy your reputation is continually to order items, usually from a distributor's catalog and through a salesman, and when they are delivered, to take a look at them and then return them—not because there is any defect in the items but because you yourself made an error in judgment or calculation. While it is true that today many firms have a restocking charge to cover this contingency, their mental computer stores the fact that you are continually returning items without cause, and they become very reluctant to fill your orders rapidly.

The location of a building yard is more or less immaterial. A great deal depends on the size of the vessel that is to be constructed. Without doubt, a suitable waterfront location is the most convenient, but it is often much more expensive than one farther inland. The differential in ongoing expenses and taxes must be balanced against the cost of transporting completed vessels to the water for launching. Also, the location should reflect the source of supplies, since the logistics of a remote location require that one purchase in large quantities with good lead times and that efficient delivery service be available. If the yard is to employ several persons, a location outside a shipbuilding area will mean that the builder must train his personnel, which again adds to his overhead. Probably the ideal location for a shipyard has a steel warehouse on one side, a marine supplier on the other, equipment manufacturers on the next street, deep water in front, and no real estate taxes.

The days of operating a business with a two-column ledger have long passed, as have the days when one paid the men at noon Saturday with envelopes containing cash. It seems unlikely that today anyone could profitably operate a small boatyard without the aid of a computer. At least monthly and preferably weekly, the computer should analyze each facet of the business, and hopefully will catch a leak before it becomes a gusher. Amounts spent for materials, supplies on hand, items back-ordered, items to be charged to each vessel under construction, payroll, payroll deductions, cost of advertising, cost of entertaining clients, office expense, accountants or legal fees, tool breakage and repair costs—all these and more must be monitored carefully. The builder with full-time employees will have to consider the following: federal income taxes, Social Security taxes, state income taxes, federal and state unemployment taxes, workmen's compensation, longshoremen's compensation, hospitalization, personal liability insurance, and product liability insurance. Books must be kept on all business transactions; records of all correspondence with clients, material orders, client billing, and so on must be maintained; and provision must be made for consultation with present and prospective clients. The most valuable associate or employee the builder can have is the person who acts as secretary, bookkeeper, purchasing agent, hostess, gofer, arbitrator—the list goes on—and that person is normally his spouse. The workmen must be paid weekly. There are legal holidays for which they must also be paid. There are maximum hours that can be worked and a minimum wage that must be paid. There are safety and health as well as sanitary features that must be considered. Pollution control, environmental impact, and wetland laws must be observed. There are building codes to be met. In some states, business or commercial licenses are required. Corporations must abide by state corporation laws and a yearly fee must be paid to the state. In fact, there are so many rules, regulations, restrictions, and other hindrances placed in the way of anyone wishing to establish a small business, not to mention a boatbuilding business, that it can become quite discouraging. The lone builder is not faced with many of these horrendous regulations, although he will have to keep books and other proof of his income and expenditures as required by law. He then may elect, when extra labor is required, to employ other individual builders or craftsmen on a subcontracted basis. He cannot set hours for the subcontractor, however, because once he does, the subcontractor becomes his employee. One interpretation says that he cannot pay by the hour but only by the job, since an hourly wage denotes an employee.

Some states have what is called an inventory tax, which means that the builder cannot stockpile material for use at a later date, thus taking advantage of temporary good prices or quantity discounts, because at the end of the year it is taxable. If the value of the material rises in the next year, it is taxed again at its adjusted value. It behooves the builder, then, to work on a basis wherein the yard owns only the unpaid portion of the vessel at any given time. In other words, all material is billed to the vessel, which is billed to the owner, who pays the builder and thus owns the material to be used for the vessel. Each month thereafter the owner is billed for any further material and all labor used, and each time he pays his bill, he acquires that much more of the vessel. In this way there is no inventory build-up in the yard at all. If the builder is building on contract and the payments are made in four or five installments, he will want to make sure he is not paying taxes on the vessel before it is completed. Also, in many states, there is a sales tax to be added to the vessel when completed, which makes the builder a tax collector. Sometimes this can be avoided by delivering the vessel to the owner in another state or out at sea. Some states have reciprocal taxing codes, and again this makes the builder a tax collector. In short, the builder can spend most of his time doing everything else except building the vessel.

Under the present laws, rules, and regulations, if I were to go back into the boatbuilding business, I would build only to my own account, use the boat for the number of days or weeks required by law, and then dispose of the vessel as a "used" boat. In this way, most of the unpleasant part of boatbuilding can be avoided. This is not to say that I would build only on speculation, but if an interested party were to appear in the yard and say, "If you had such and such type of vessel and it were completed at such and such a time, I would definitely be interested in buying it," there would be nothing to prevent him from lending me the money to build the boat, which would theoretically commit him to purchasing the boat upon completion. Legally, he must charge interest on that loan, which of course would be added into the cost of the vessel. It takes a bit of calculation to work all of this out profitably for both sides.

There are three types of boatbuilders. There is the innovative individual who builds to the best of his ability. He charges what it costs to build, remains competitive, and hopefully makes a profit on his investment. He will end his days well satisfied with his accomplishments, but probably no richer than the day he started. There is the individual who overcharges, gouges without flinching, does slipshod work, and is a hog about everything. A hog is destined to be slaughtered, and he does not last long in the business. The third type is the follower of leaders. Nothing innovative ever comes out of his yard. He merely copies what everybody else is doing, and then, in order to be competitive, he cuts corners and charges less for his product. This type is a sheep, and a sheep is destined to be shorn at least twice a year.

Regardless of who is paying for it, a vessel under construction is the builder's vessel, his creation, his sweat, his worry, his headache, and his pleasure. The best sight in the world is the transom of a finished vessel heading out to sea, which means to the builder that he can start another one. The next best sight is her return with a completely satisfied owner aboard.

10

▽

VESSELS
IN
COMMERCE

Other than piracy, which is frowned upon, there are numerous occupations that a vessel may pursue. Indeed, a good living can be earned under the right circumstances, and historically, fortunes have been won in properly managed vessels. The age-old question when designing, building, or financing a new vessel is, "Will she pay?" To this question one must answer, "If properly designed, built, sailed, and managed, YES!" (I apply the word "sail" indiscriminately to both sailing and motor vessels, except when it is obvious that the term applies only to a sailing vessel.)

Any vessel built strictly for commercial freight purposes requires an individualistic owner-master, for there is nothing that society and government agencies fear more than the individual entrepreneur, who demands and seeks free enterprise and will not fit into the mold of conformity. He is a thorn in their sides. To be successful, one must first be a competent seaman. In fact, one of the Rules of the Road (Article 29, old version) is as follows: "Nothing in these rules shall exonerate any vessel, or the owner or master or crew thereof, from the consequences of any neglect to carry lights or signals, or of any neglect to keep a proper lookout, or of the neglect of any precaution which may be required by the ordinary practice of seamen, or by the special circumstances of the case." The implication of this rule is all-encompassing; there is no loophole. The mere fact that a person can afford to purchase a vessel does not excuse him from being a competent seaman, and he *must* possess that expertise required "by the ordinary practice of seamen." "Ordinary practice" is defined by the collective sense of what experienced and competent seamen would do in specific circumstances—not just what Joe or Bill might do. By law, a seaman must be tried by his peers, and this produces a harsh jury.

Seamanship cannot be learned from a book. It can only be learned from experience at sea in all types of weather and other conditions found only at sea. Seamanship is not just steering, tying knots, or navigating. It is all things that pertain to the management and maintenance of the vessel. Thus, before embarking on a venture in a commercial vessel, one would do well to gain experience at sea in other commercial vessels in a similar trade. Then, it is incumbent upon a seaman to know his own vessel and to know the sea, but most of all, to know himself. Decisions made in handling a vessel must be natural, instantaneous, and correct. There is no time to ponder, have a quarterdeck conference, or hold a democratic vote.

Second, one must know his secondary trade as well as he does his seamanship. To engage in fishing, he must be a fisherman, an occupation that one can learn well by having been born into it. Like seamanship, fishing cannot be learned from a book. To carry freight requires a complete knowledge of maritime law and the civil law pertaining to the trading area, as well as weights and measures, port regulations and conditions, currencies, and the economics of the trade. To work one's vessel as a trader requires all the knowledge of the freighter plus the ability not only to buy but also to sell one's cargo. A profit must be made or the owner will soon be out of business. The "perils" of the sea demand that the vessel be properly designed, built, and suitable for her employment, and also properly maintained, stowed, navigated, and manned.

Last, but of equal importance, is the selection of the correct vessel for the chosen trade. It is usually wise not to choose a vessel so specialized and restricted in her design or construction that she is rendered unsuitable for any other purpose at a later time in her life.

The demise of small freight-carrying vessels in the United States was hastened by the construction of numerous roads and highways. This brought about an increase in trucking freight, which, to an extent, was subsidized by the taxpayers paying for and maintaining the roads. Furthermore, the freight rates and routes were regulated by the Interstate Commerce Commission, which granted franchises, thereby stifling any competition.

I remember paying $25 sea freight on an engine brought from Norway to Norfolk, Virginia, and then having to pay $75 freight to truck that engine from Norfolk to my yard only 65 miles away. It could not be delivered direct from Norfolk, but first had to go to Richmond, Virginia, which is 75 miles from Norfolk, and then be transshipped another 75 miles to the yard. The actual freight rate and route was determined by a franchise granted by the ICC. It was often cheaper to use a boat to transport or pick up items in Norfolk, Newport News, Hampton, or Baltimore than it was to become involved in truck transportation. Indeed, in the 1950s, I would order marine supplies from Baltimore by telephone on one day and meet the Chesapeake Bay Steamer the next morning in Hampton to pick up the supplies. Until the 1970s, a small coastal tanker used to discharge oil and gasoline at a dock directly across the river from my yard. There are still areas where water routes are more direct than land routes, but renewed activity by small vessels will become possible only when those in power realize that waterborne transportation is often less expensive than truck and rail transportation, and that the cost of maintaining waterways is cheaper than building and maintaining highways and bridges. There is no reason why all three modes of transportation should not exist together, according to the economics of each.

The building of large freighters and tankers is federally subsidized with construction cost differentials to offset the higher costs of building in the U.S. versus building foreign. There is also a bonus for incorporating "defense features," and there is protection against foreign competition in the coastwise trades along with regulated freight rates. The small freighters and tankers, however, have to be built with the owner's own capital and without help from anyone. The small vessels encountered restrictions, regulations, and other hindrances which made them uneconomical to build, own, maintain, and man, and they were driven to oblivion. This resulted in many of the old, privately owned and maintained wharves being abandoned or allowed to rot away, and under our present laws, these would be difficult to replace.

The same is true in most of the island trade. In the not-too-distant past, a great deal of in-island as well as inter-island trade existed. A vessel would receive cargo at numerous ports along the coast and then transship it to that island's principal port or to another island, where it would go into a warehouse, a processing plant, or the hold of a waiting steamer. The demise of small vessels in the islands occurred over a period of about a decade. As each island nation became independent, it seemed that its first assertive act was to build roads, airports, and government buildings that really did not add to the economy at all. Trucks would be imported to do what small boats had always done. Fuel for the trucks would have to be imported and stored, which further drained financial resources. The small landings and wharves would be abandoned and allowed to deteriorate. The need for casual labor to load and unload the vessels would cease, and employment would decrease. The way of life would not become better, and the trucks would soon destroy the roads. Proper maintenance of the roads was an expense these islands could ill afford. Contrary to the situation in the United States, there would be no way to levy more taxes for the highways, especially if the population were virtually unemployed. So the roads would develop potholes, which would become craters, and within a few years they would become almost impassable. Then produce could no longer be transported economically by land, but by this time, the small vessels would have been sold or would have deteriorated beyond repair. The few that had been able to eke out a living would begin to enjoy greater revenues, and this would lead to a resurgence of building new vessels for the inter-island trade.

While it is true that most island nations protect their citizens with laws that restrict competition from outsiders, there is still room for an individual to pick up cargoes, especially if he is willing to transport greater distances. In the Caribbean, over 100 types of cargo are transported by small vessels. They may be broken down into several categories:

● *Spices, herbs, roots, barks, and gums.* Allspice; cassareep; copra; arrowroot; ginger; Dasheens; guaiac; aloes; banana powder; Canella bark; Tumeric; cinnamon; Chilices; manioc; Malangas; Nispero; cascarilla bark; and others.

● *Seeds and beans.* Coffee; divi-divi; castor; aniseed; cardamom seeds; annatto; cocoa; sesame; and others.

● *Nuts.* Coconuts; cashew; kola; Brazil; Castanhas; coquilla; palm; and others.

● *General.* Bones; cattle hair, briquettes; sugar; salt; Candelilla; beeswax and bayberry wax; dyewoods such as fustic; other logwoods; lignum vitae; orange peels; boxwood; etc.

● *Building materials.* Bricks; asphalt tiles; ceramic tiles; cement; shingles; pipe; bathroom fixtures; windows; doors; paint; etc.

● *Miscellaneous.* Auto tires; bottled drinks; canned drinks; bottled rum; tins of food; bags of potatoes and onions; empty glass and plastic bottles; tins of biscuits; fertilizers; etc.

● *Deck cargoes.* Cars; trucks; small boats; tractors; bicycles; bedsprings; empty (MT) drums; asphalt in drums; oil in drums; lumber; creatures; poultry; passengers; etc.

Logwood stows at 140 to 190 cubic feet per ton, but can be used for "chocking" and stowed within other hollow cargoes such as tires, thus lowering the stowage factor of a mixed cargo. One cargo transported to an island with a population of just a few thousand might be a four- or five-year supply of that product. Therefore, vessels need not be gigantic for this type of trade. A stowage factor of 50 to 60 cubic feet per ton is acceptable and economical for a vessel, especially if her hull shape permits the carriage of cargo on deck. In the inter-island trade, vessels of 40 to 60 feet on deck seem to be the best size to own and maintain, work only parttime, and still show a profit at year's end.

Sometimes a small cargo vessel can exist only where the amount of freight to be carried is so small, or the port from which it is loaded so remote, that a larger vessel would find it uneconomical to pick up. It has been said that a small business can only exist in the trough of big ones. This is apropos to the small vessel, for, when and if she becomes too active or tries to expand, it seems that someone is always ready to jump in and try to drive her out of business.

The U.S. coastal freight commodities are numerous, and include grain, feed, fertilizer, bricks, cement, lumber, cordwood, hay, fresh fruits and vegetables, and building materials. On the U.S. coast, a stowage factor of 70 cubic feet per ton seems to be a good all-around average for general freight. If a specific cargo is assured, such as cement in bags at 27 cubic feet per ton, or grains at 45 cubic feet per ton, or fresh foods that average out at 120 cubic feet per ton, then the volume of the vessel's hold can be designed around the cargo. The following general measures apply:

a cubic yard is 27 cubic feet;
a cubic meter is 35.31 cubic feet;
a cord of wood measures 4 feet by 4 feet by 8 feet, or 128 cubic feet;
a shipping ton is 40 cubic feet of merchandise;
a shipping ton is 42 cubic feet of timber.

The following timber designations apply:

> 100 board feet = 1 square
> 120 deals = a "hundred" deals
> 108 cubic feet (3 feet x 3 feet x 12 feet) = 1 stack (half a fathom)
> 216 cubic feet (6 feet x 6 feet x 6 feet) = 1 fathom
> 50 cubic feet of square lumber = 1 load
> 40 cubic feet of unhewn lumber = 1 load
> 600 square feet of 1-inch boards = 1 load
> 400 square feet of 1½-inch boards = 1 load
> 300 square feet of 2-inch boards = 1 load
> 200 square feet of 3-inch deals = 1 load
> 150 square feet of 4-inch deals = 1 load
> 218 running feet of 3-inch x 11-inch planks = 1 load
> 267 running feet of 3-inch x 9-inch planks = 1 load
> 34 running feet of 3-inch x 7-inch battens = 1 load
> Pieces of timber up to 7 inches wide are "battens."
> Pieces of timber 8 to 11 inches wide by 2 inches or thicker are "deals."
> Pieces of timber 11 or more inches by 2 inches or thicker are "planks."

In small vessels, the actual usable cubic is often lessened by the shape of the hold. While freight may be laid alongside based on timber or merchandise tons, the actual conversion to capacity tons must be modified by the stowage factor times the usable volume of the vessel to determine if there will be a profit on that particular cargo. Vessels carrying deckloads of lumber like to be full-ended with a generous beam.

Basically, if the vessel is not loaded to her full cubic but is loaded to her loaded draft, space is sold as weight. If the full cubic is used but the vessel is not loaded to her marks, then space is sold by the cubic. The maximum earning capacity is achieved when the holds are full and the vessel is to her marks at the same time. Therefore, the shape of the hull is related to her cargo. The flat-floored motor vessels have the advantage over the steeper-floored sailing vessels in that little or no dunnage is needed in the former to "chock off" the cargo, and draft is less for the same amount of cargo.

Barrels of liquid are usually carried on the deck of small vessels, since the holds (of sailing vessels in particular) severely limit the number that can be stowed. (Barrels cannot rest against the bilges or sides of the vessel.) Steel drums are always stowed end up, thus lending themselves even more to deck stowage. Empty (MT) drums are an excellent deck load. Case oil is carried below decks but must be well "chocked off" to be kept level, square, and fully bearing; otherwise, damage and leaking will occur.

Some other figures that determine hold and deck layout and capacity, especially for inter-island trading, are:

> A quarter barrel (firkin) = 9 gallons; an anker = 10 gallons.
> A half barrel (kilderkin or rundlet) = 18 gallons.

A barrel = 36 gallons.

A tierce = 42 gallons.

A hogshead varies from 46 to 57 gallons and should be measured before accepting unless the original contents are known.

An imperial gallon weighs 10 pounds.

Ice weighs 57 pounds per cubic foot.

Bulk grains usually stow at 45 to 50 cubic feet per ton.

In the U.S. coastal trade, the usual measure is a short ton, or 2,000 pounds. Elsewhere, it is the long ton, or 2,240 pounds.

In the contract for delivery, there is often reference to laying the vessel or landing a cargo within a specified number of cable lengths of a specified destination. A cable is 100 fathoms, 600 feet, 200 yards, or 183 meters. The charts should be checked to ascertain whether the vessel can lay that close, and the terrain must be checked if the cargo is to be landed on the beach or a position above the beach.

The metric system has become more or less universal for the calculation of measures, but in many of the small ports of the world, old measures are used:

In Argentina, 1 frasco = 2½ quarts; 1 baril = 20$\frac{1}{10}$ gallons; 1 libra = 1 pound; 1 vara = 34$\frac{1}{10}$ inches; 1 arroba = 25.3 pounds (dry); 1 quintal = 101.4 pounds.

In Brazil, 1 arroba = 32.4 pounds; 1 quintal = 130 pounds.

In Chile, 1 fanega = 2.5 bushels (dry); 1 vara = 33.3 inches.

In Mexico, 1 vara = 33.4 inches; 1 carga = 300 pounds; 1 fanega = 1.6 bushels (dry); 1 arroba = 4.3 gallons.

It can be seen that the same terms may connote different quantities from port to port.

I have made repeated reference to "pure" sailing vessels versus auxiliary sailing vessels and motor vessels. Under 700 gross tons, a pure sailing vessel does not have to have a licensed master or others in her crew. A vessel of this size would be well above the 79-foot limit set forth in this book (79 feet being the maximum length not requiring a Load Line Certificate). Thus, a pure sailing vessel may carry freight for hire without being regulated, except for the carriage of dangerous cargoes and flammables in bulk. It may also be used as a research vessel, and it can carry six or less passengers for hire without regulations.

A vessel with a propelling engine is, technically, a motorboat if it is less than 65 feet in length, or a motor vessel if it is over 65 feet in length; it makes no difference if the vessel also carries a full sailing rig. A motor is a motor, and its horsepower is immaterial. A motorboat of over 15 gross tons cannot carry freight for hire without meeting further regulations. In general, all motor vessels over 300 gross tons must be certificated and inspected by the U.S. Coast Guard.

A motorboat carrying six or less passengers for hire does not need to be inspected or certificated, but her operator must hold a Motorboat Operator's License for the appropriate route. (This license allows a person to operate motorboats or other vessels of 15 gross tons or less.) A vessel carrying more than six passengers for hire comes under the inspection laws.

A vessel of more than 100 gross tons carrying more than six passengers for hire is inspected under Subchapter H of the U.S. Coast Guard regulations, and a vessel of less than 100 gross tons carrying more than six passengers for hire is inspected under Subchapter T.

Vessels that must be inspected and certified under one or more of the subchapters also require plans approval by the U.S. Coast Guard, both for new construction and for significant alterations. The additional cost of preparing these plans and meeting the building requirements for certification add significantly to the final cost of the vessel.

My personal belief, therefore, is that a pure sailing vessel with no auxiliary gives one the most freedom from regulations and restrictions; a motorized yawlboat could be used, since this would not alter the vessel's status. If a motor is incorporated into a vessel, she should be less than 100 gross tons and 65 feet and should carry no more than six passengers for hire in order to be subject to the minimum regulations. If the vessel is to be used for freighting, she should be less than 15 gross tons.

The size of a vessel is important in many ways. The gross tonnage determines which category the vessel will fit. The net tonnage determines what her harbor and light dues will be, and it behooves the owner (and thus the designer) to obtain the smallest net tonnage possible. A properly designed sailing vessel can double her gross tonnage with the deadweight tonnage that she will carry.

Under a recent change in the laws, a vessel's document, issued by the U.S. Coast Guard (Department of Transportation), indicates all the categories in which the vessel is permitted to operate; the trade in which she is engaged at a particular moment is the one that regulates her customs procedures. The yearly renewal is a sticker that the master or owner of the vessel receives by mail and pastes to the document. For the small vessel, which may be three or four years away from a U.S. port while tramping cargoes, the biggest problem is mail, which often catches up with her two or three months late or may be lost altogether. At such time, if her document has expired, she is operating illegally. The old document made allowance for this contingency, but the new document, for its part, has eliminated much of the paperwork that formerly was required when a vessel shifted from fishing to coastal trade or to foreign trade and back again. At one time, it was not at all unusual for a U.S. vessel, at the end of her fishing season, to carry a coastwise freight (say of lumber or cordwood), discharge it, and then load for a foreign port; discharge that cargo and load back to a U.S. port; discharge cargo again and pick up another coastwise freight; deliver that freight and then refit for another fishing season. If there can be any objection to the Customs Service, it is not with their personnel, whom I, at least, have always found most helpful and courteous, but with the locations of many customs houses, which are seldom convenient to the docks that the small vessels frequent. One must rent a car or go by taxi in order to enter and clear his vessel.

Yachts are often engaged in trading voyages to the owner's account, which means he owns the cargo and maintains the right to sell it. If trading is the primary purpose of the vessel, then this is the simplest solution—in other words, to remain classified as a yacht. The disadvantage of not being documented commercially is that the vessel cannot accept freight or passengers when they are available. The yacht cannot carry passengers for hire or be chartered other than in a bareboat capacity. The owner does have less paperwork when using his vessel strictly as a yacht and not for a trading voyage.

One myth holds that a yacht does not need clearance papers when going foreign. True, a yacht need not have clearance papers to leave U.S. ports, but clearance is definitely required in order to enter most foreign ports. In many instances, stiff fines are levied upon vessels failing to meet this requirement. It should be mentioned that there is no way that a yacht can be given clearance from U.S. ports, so there are only two ways for her to be legal upon arrival in another country. She may sail first to a country that does not require a clearance from the U.S.—such as the Bahamas, Bermuda, Canada, and other countries that maintain reciprocal yachting privileges with the United States—and there obtain a clearance to the next country to be visited. The other and best way is to obtain a personal consular visa and a visa for the yacht for each country to be visited; however, a visa can be obtained for just one country if the vessel is to be sailed direct to that country. Upon arrival there, one encounters no problem in obtaining a clearance to another country, and the vessel will be in the mainstream of entering and clearing procedures.

When a vessel is bulk or charter freighting, agents that are employed in most ports can handle the details of entering, clearing, and the posting of bonds for dutiable items. In many foreign ports, much business is done by relatives of the powers-that-be, and the lay days required in order to obtain a berth for discharging and loading freight can be lessened by having the right agent and paying a little "mordita." Tied up to a seawall or wharf or in the careenage, the master often just hangs a sign in the rigging stating that freight is being accepted for specified destinations. Since the waterfront is the center of island activity, this method can work quite well. In the larger ports, one can often obtain freight by posting a notice in the local newspaper.

Insurance of the cargo is another matter that must be resolved. It is often so expensive that there remains no profit from the freight. In such cases the owner must insure the cargo himself, which requires capital. The method most often used is to assign funds in one's own bank payable to either the consignee or consignor, as the case may be, upon failure of delivery. This does not entail much risk with cargoes of little value, but cargoes such as rum, spices, and herbs are so valuable that they can equal or exceed the value of the vessel. Ordinarily, when the master of a vessel is known for his integrity, he is not refused cargo because of a lack of insurance. It is best, of course, if the cargo can be underwritten for a nominal charge. The small inter-island freighters are seldom insured for the loss of the vessel.

Smaller vessels have the advantage of not requiring a large crew. The paperwork for alien crew aboard U.S. vessels is voluminous, and if the paperwork is not in order upon arrival in the U.S., the alien is usually detained by the Immigration and Naturalization Service, delivered to the nearest airport, and shipped out of the country on the first plane, all at the expense of the owner of the vessel that brought the alien into the country. In most instances, the master of the vessel must also post bond for each alien. So, if one is sailing under the U.S. flag, it is best always to have an American crew or one that has valid visas for the United States.

At first consideration, sailing under a flag of convenience seems an attractive option, since one may eliminate many of the restrictions placed on owning and sailing an American flag vessel. Most countries that offer a flag of convenience require a fee of some sort or incorporation in that country, which involves another fee, and so on. If one wishes to sail under another flag, it helps if he is fluent in the language of that country, and it certainly helps if he is married to a citizen of that country. Sometimes the law requires that ownership of the vessel be controlled by a national, with the real owner retaining only about 45 percent. Then his "partner" could possibly demand to be bought out even though his ownership is in name only. Finally, the vessel that elects to transfer to another flag but later wishes to return to the U.S. flag loses all her rights to trade coastwise; she is considered to be a foreign vessel forever, despite the fact that she was constructed in the United States.

So, in conclusion, as the economies of all nations begin to shrink and the standard of living contracts, many of the larger vessels will become uneconomical and will have to be laid up. Smaller vessels will take up the slack. While falling oil prices usually mean cheaper fuel in the world markets, these savings may not be passed along to the consumer. In highly industrialized countries and those with oil reserves, the rise in oil prices in the 1970s was viewed as a continuing fact of life that would never alter, and they were eager to lend large sums of money to other nations in return for raw materials. Whatever goes up must come down. In the future, this artificial manipulation of economies will come home to roost.

The employment of a vessel in the commercial trades requires much thought by the owner, designer, and builder, who, working in harmony, can produce an excellent vessel that is well suited to her trade. The success of the small vessel lies in the tenacity and acumen of the owner. One should always remember that there is honey in horse manure, but it takes a bee to find it.

$\triangledown$

AFTERWORD

The first-time builder who plans to construct a yacht, fishing vessel, or commercial vessel should find sufficient information in the two volumes of *Steel Boatbuilding* to produce a proper vessel, ready for sea. If an owner or builder is converting a vessel from one service to another or merely improving it, Volume 2 will help him over the rough spots. Should you decide to enter the boatbuilding business, then I say "Welcome!", and should you decide to use your vessel commercially, I hope you will always sail with a profit.

The professional builder, I hope, will also find these volumes useful, and a good reference for some of the oddities that can occur in the "business."

I have used the methods and ideas set forth in these volumes successfully for many years and numerous types of vessels. They may not be the best ways, and they are certainly not the only ways, but they are good ways. However, I retain the right to change my mind and use different methods for the next vessel I build.

To one and all who build their vessels and venture to sea, my wish is:

"May fair winds escort you!"

$\triangledown$

GLOSSARY

Air Course. An air passage or channel.

Babbitt. A white antifriction alloy composed of copper, antimony, and varying proportions of tin.

Ballast Tanks. Tanks devoted to carrying water exclusively in order to bring the vessel down to a usable waterline when light. These tanks cannot carry fuel, potable water, etc., without being measured as part of the net tonnage.

Becket. A rope grommet or metal eye at the bottom of a block, to which the standing end of a fall is secured.

Bight. A bend or loop in a rope.

Block. A fitting consisting of one or more grooved pulleys or sheaves mounted in a casing or shell fitted with a hook, eye, or strap by which the block may be attached.

Bolster. A plate or shape at the outboard end of the hawsepipe.

Boom Jaws. End fittings made of wood or metal that partially encircle the mast to keep the boom in place; they replace a gooseneck.

Bosun's Stores. All small fittings, spare parts, rope, and other small items necessary for the maintenance of the hull and rig. In small vessels, this also includes the contents of the paint and rope lockers.

Brigantine. A two-masted vessel that is square-rigged on the foremast.

Bulkhead. A vertical partition separating different compartments or spaces from one another, which may be sited transversely or longitudinally.

Calcium Chloride. A crystalline compound used as a drying agent.

Chamfer. To bevel the edge of a plate or plank.

Chocking Material. 1. Miscellaneous pieces of wood or timber used in conjunction with other dunnaging material to brace cargo and prevent it from moving. **2.** An epoxy material of pouring consistency that is used in lieu of other materials to maintain the alignment of machinery on its foundation (bed, seat, or platform).

Clad Steel. The result of a process wherein a thinner layer of another metal is bonded to a thicker layer of steel, either by adhesives or by explosives. In marine use the most common claddings are stainless steel, aluminum, and cupronickel. Two different welding processes are required when working with clad steels. Other metals may also be bonded, such as lead, silver, gold, zinc, and so forth.

Clevis. A U-shaped piece of metal with a hole worked into each end to receive a pin.

Club. A small spar laced to and named after a sail—for example, jib club.

Coleman Hook. A split hook, one-half of which has a lip with a hole in it, while the other has a slot that receives the lip. A leather thong is run through the hole to secure the two parts.

Cribbing. Wooden shores or blocks placed under a vessel's keel and bilges in a building berth or cradle.

Deadeye. A stout disk of hardwood strapped with rope or iron, through which three holes are pierced for the reception of lanyards. Deadeyes are usually made of lignum vitae and are used as blocks to connect shrouds and chainplates.

Deadweight Cargo Capacity. The weight of the vessel with everything in her when loaded to a given waterline. It is expressed in pounds, tons, or cubic feet.

Declivity. A gradual descent or slope.

Declivity Board. A device used in conjunction with a carpenter's level for plumbing vertical members and for leveling members that run in the fore-and-aft direction of a vessel on the building ways. It compensates for the slope of the ways, and may also be used when building a vessel's interior afloat.

Dinghy. A small boat used as an extra boat or tender for a vessel.

Dolly Bar. A tool used to turn or hold angles, tees, or similar bars while they are tack-welded, riveted, or fitted to other objects.

Drag. The designed excess of draft aft in a vessel when compared to the draft forward with the vessel floating at her designed waterline.

Escutcheon Pin. A round-headed pin originally used to put an escutcheon plate (or keyhole plate) onto a door.

Faying Surface. The meeting surfaces of two pieces of wood or metal that are bolted, riveted, or welded together.

Fiddles. A rail made of wood or rope around the edges of a counter or table to keep objects from falling off when heeled or in rough weather.

Forepeak. The space between the collision bulkhead and the stem.

Gooseneck. A fitting that secures the hinged end of a boom to the mast, allowing it to move up and down and athwartships.

In Ballast. A term that describes a vessel without cargo.

Jackstay. A stay that does not go to the rail but instead attaches to the mast and to the deck and has a sliding fitting by means of which a yard or sail can be raised.

Jibheaded Sail. A fore-and-aft sail of triangular shape, fitted to the after side of a mast; also known as a marconi, Bermudian, or leg-o'-mutton sail.

Jury Rig. A temporary rig.

Kentledge. Pig-iron ballast; scrap iron.

Ketch. A vessel with two masts, the aftermost of which is shorter and forward of the rudderpost.

Knockabout-Rigged. A term describing vessels that lack a bowsprit.

Lanyard. A rope rove through deadeyes and used for setting up the rigging. The dead end is fitted through a hole in the upper deadeye, and a double Matthew Walker knot is worked into the end. After setting up the lanyard, its tail is hitched around the upper deadeye at the shroud with a cow hitch.

Lazarette. The aftermost compartment of a vessel.

Lazyjacks. Lines looped under the boom through eyes. They prevent the sail from falling on deck when lowered.

Limber Chain. Chain that is led through the limber holes in a vessel and is pulled from one or both ends in order to clear any debris that might be blocking the flow of water through the holes.

Luting. A mixture traditionally of putty, white lead, and linseed oil, having the consistency of heavy cream. It is used by boatbuilders to ensure watertightness of joints. Today, it is usually a polysulfide compound.

Magnesium. A silver-white metallic element, malleable, ductile, and light.

Maul. A heavy mallet or hammer.

Micarta. A plastic material consisting of layers of a fibrous product impregnated with synthetic resin and then hotpressed to achieve great density and high mechanical strength. It is used for stern tube and propeller strut bearings and for rudder pintle bushings.

Mortise. A cavity or hole into or through which some other part fits or passes. A cavity cut to receive a tenon.

Muscovite. Potash, mica.

Mylar. A proprietary polyester film.

Parcel. To protect rope from the weather by winding strips of canvas or other material around it before serving.

Parrel. A rope loop or sliding collar by means of which a yard or spar is held to the mast.

Pay. To coat or cover with tar, pitch, or bituminous cement, or to fill the caulked seams of a deck with hot pitch or marine glue. Today, two-part polysulfides are used.

Pipe Berth. A berth made of pipe with canvas stretched over it or suspended from it as a mattress. The berth is usually made to hinge upward and to be easily removed.

Plenum. The opposite of vacuum. The word denotes fullness, or a condition in which the pressure of the air in an enclosed space is greater than that of the outside atmosphere.

Porthole. An opening in the side or superstructure of a vessel to give light and ventilation to the living quarters.

Portlight. Another name for a porthole. In its proper usage, portlight would connote a nonopening porthole; however, the two words are used interchangeably today.

Poultice Corrosion. A soft spot in metal forming when two surfaces that are in supposed contact with each other are in reality separated by a gap in which moisture and air have collected.

Prismatic Coefficient. A coefficient denoting the volume of a hull relative to a prism having the same length as the hull and the cross-sectional shape and dimensions of the hull's midship section.

Pushpit. A railing at the stern of a yacht and similar in its construction to a pulpit or bow rail.

Rail Rollers. Wooden or metal rollers one to several feet long, mounted on axles and suspended just above the rail. On fishing vessels they are used for bringing the nets and other gear over the rail. On vessels that load deck lumber, they are often used to slide the lumber onto or off the deck. If a vessel's rig has a limited lifting capacity, they are used for sliding boats onto and off the deck.

Ringtails. Extensions of the gaff and boom to which one or more cloths of sail are stretched, adding sail area.

Riprap. Scrap stone.

Rudder Port. A hole through the hull for the rudderpost. In steel vessels, it is a watertight pipe or casting. If stopped below the waterline, it is fitted with a gland or stuffing box; if carried through to the main deck, it is fitted with a bearing.

Schooner. A fore-and-aft rigged vessel with two to seven masts. In a two-masted schooner, the foremast is the shorter of the two. In a schooner having three or more masts, each mast after the foremast is usually one or two feet higher than the one just forward of it; there are many exceptions, however, especially in three- and four-masted schooners, where the after mast can be shorter even than the foremast.

Seizing. The binding or whipping of a rope or ropes using small stuff.

Serve. To bind or wind a rope tightly with small cord or marline, keeping the turns very close together.

Sheave. A grooved wheel in a block, over which a rope, wire, or chain passes; a pulley.

Sister Hook. Two hooks that come together to form an eye. They look like the top halves of two question marks facing each other.

Skeg. The after part of the keel of a vessel, near the sternpost.

Small Stuff. The general name given to all the small lines or ropes used on board ship.

Spectacle Iron. A steel fitting used at the clews of squaresails or on the headstay of some vessels showing a Chesapeake Bay influence. In the former case the fitting has three eyes, the outside eyes being attached to the sail's head rope and leech rope and the center eye being attached to the clew outhaul or earing. In the latter case, this iron fitting contains three holes, of which the center one, being the smallest, is of the correct diameter to fit around the headstay. The fitting is suspended from the masthead at the desired location by means of a bridle. To the outer, larger holes are attached the jib lazyjacks. The spectacle iron is never fitted to the jibstay.

Spring Stay. A horizontal stay extending between the lower mastheads.

Studding Sails. Light sails set as extensions off the ends of a yard.

Swage. To extrude.

Tang. A metal fitting used on sailing vessels for the attachment of standing rigging to masts.

Tenon. A projected member left after cutting away the peripheral wood. The tenon is inserted into a mortise to make a joint.

Thumb. A cleat with one horn used on a mast to prevent the rigging from sliding. It is also used elsewhere on a vessel to alter the lead of running rigging.

Tie Coat. A coat of paint that forms a bond between two mutually incompatible paints. The tie coat paint must be compatible with both.

Topgallant Sail. A squaresail set above a topsail on a mast above the topmast.

Triatic Stay. A wire leading from the lower masthead of one mast to the topmast head of the next.

Tumblehome. The slope of a surface, such as a topside, that inclines inward and upward toward the vessel's centerline.

VCG. Vertical center of gravity.

Water Sail. A sail that fits under a boom or the bowsprit to add to sail area.

Working Boat. A small boat carried aboard a vessel for the loading and discharging of cargo and for other heavy work. This boat may also be used as a dinghy on small vessels.

Worm. To fill the contline between the strands of a rope with tarred small stuff or filling to prevent the penetration of moisture to the rope's interior.

Yawlboat. A boat hung in davits over the stern of a vessel. Such a boat is often equipped with an inboard engine for service to a large sailing vessel, or an outboard for service to a small sailing vessel. A yawlboat is used to push or help maneuver the vessel in rivers and harbors. In some localities, it is called a push boat.

$\bigtriangledown$

INDEX